ALL ABOUT
PLUMBING
and Central Heating
ROY DAY

HAMLYN

London · New York · Sydney · Toronto

Published by
The Hamlyn Publishing Group Ltd
London · New York · Sydney · Toronto
Astronaut House, Feltham, London, England
© Copyright The Hamlyn Publishing
Group Limited, 1976
ISBN 0 600 30254 7
Set in 9 on 11pt Monophoto Plantin by
Tradespools Ltd, Frome, Somerset
Printed in England by
Chapel River Press (IPC Printers) Ltd

Acknowledgements

The publishers would like to thank the following for help with
photographs for this book.

Armitage Shanks; Barking Brassware, Ltd.; Belling & Co. Ltd.; F. H.
Bourner & Co. Ltd.; Evans and Graham (Heating) Co. Ltd.; Fordham
Pressings, Ltd.; The Finnish Valve Co. Ltd.; Fibreglass Ltd.;
Fibrewarm Products Ltd.; Gainborough Electrical Ltd.; GEC-EXPELAIR,
Ltd.; William Heaton & Co. Ltd.; Miraflow, Ltd.; Peglar-Hattersley,
Ltd.; Prestex; Pifco; Summerhill Heating Services, Ltd.; Walker
Crossweller & Co. Ltd.

Line drawings by Bob Mathias

ALL ABOUT
PLUMBING
and Central Heating
ROY DAY

Contents

Introduction 6

Water Supply 7

Ball-Valves 16

Hot Water Systems 21

Toilet Cistern Troubles 26

Taps Old and New 30

Pipes and Joints 41

Frost Precautions 50

Clearing a Blockage 57

Outside Taps 70

Scale and Corrosion 74

Condensation 78

Showers, Baths, Basins, W.C.s 80

Central Heating 100

Improvement Grants 111

Plumbing Terms 111

Index 112

Introduction

For far too long the mechanics of plumbing were a closed book to the ordinary householder. Most of the work involved was considered by many to be beyond a layman's capabilities.

All that has changed. Now a great army of enthusiastic do-it-yourselfers regularly carry out plumbing jobs in their homes – from renewing the humble washer to installing a complex central heating system.

But there is still another large number of householders to whom plumbing remains a mystery. It is for them that this book has been prepared.

A good grasp of how your plumbing system works is the first step to carrying out your own repairs and thus helping the family budget. My aim is to describe average plumbing systems in terms easy to understand, and to explain how simple plumbing jobs can be done. The book is not intended as a manual for students.

Roy Day

Water Supply

Some domestic plumbing arrangements are more complicated than others but, generally speaking, basic layouts follow a traditional pattern.

One of the most important things to know is where vital stopcocks and stopvalves are positioned. Even if you are not prepared to make minor plumbing repairs yourself, it is essential to know which stopcock or stop-valve to turn off in the event of an emergency.

Some parts of a plumbing installation are not visible. For example, take the stopcock of the pipe which carries the water into the house. This is often buried under the pavement outside the house – but not always. It could be under the footpath which leads to the front door or, in some old houses, it could be buried somewhere in the garden. Normally, you will find the stopcock under the pavement.

INTERNAL STOP-VALVE

In modern houses it should not normally be necessary for a householder to use this stopcock. Usually the pipe which supplies the water to the house has a stop-valve fitted *inside* the house, as a rule near an entrance. Many older houses, however, lack this internal refinement, so to turn off the water the occupants have to rely upon the outside stopcock.

Even if you have a stop-valve inside the house, it is obviously an advantage to be able to turn off the water from the outside stopcock. This applies particularly when the inside stop-valve needs a new washer.

Few houses, comparatively speaking, have a key to operate the outside stopcock and they are not always easy to obtain.

The stopcock is usually buried about 30 inches below ground level, *fig. 1*. This is to protect it from frost. These stopcocks normally close in a clockwise direction. They are sometimes enclosed in a box or a length of drainpipe.

There is a hinged cover at the top at ground level which should at all times be kept clear of soil, weeds and anything else which may make the cover

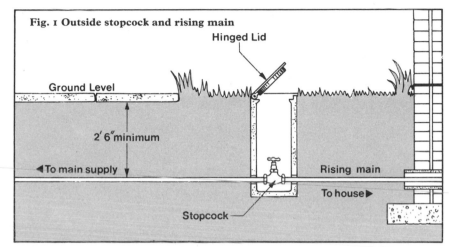

Fig. 1 Outside stopcock and rising main
Hinged Lid
Ground Level
2' 6" minimum
◄To main supply
Rising main
To house►
Stopcock

difficult to open. Keep the hinged flap clean and put a spot of oil on it occasionally.

DRAINCOCKS

The stop-valve inside the house, *fig. 2*, should have a draincock fitted near it, *fig. 3*. In most cases both stop-valve and draincock are combined in one fitting, *fig. 4*.

A draincock will have a hose connection so that the rising main can be drained if it has to be repaired, or as a frost precaution if the property is to be left empty during the winter months.

The stop-valve should turn freely at all times. So it is a good idea to make a regular check that it will do so. Turn the water on and off two or three times when you make this check.

Fortunately, the washer on this stop-valve does not wear out very often. However, they do not last for ever so you will need to know how to rewasher your valve. For this see the section on rewashering taps.

RISING MAIN

The pipe which carries the water into the house is called the rising main. It climbs up through the house and supplies the cold water storage cistern normally situated in the loft. This pipe should at all times be protected against frost (see chapter on frost precautions).

The water in the storage cistern should never be used for drinking

purposes. This is because it can easily become contaminated by insects crawling into the cistern. Drinking water should come direct from the main – that is, normally, from the kitchen tap. The diameter of the rising main pipe is ½ in.

After supplying the kitchen tap with water, the rising main pipe normally continues its journey through the house up into the loft. There it feeds the cold water storage cistern, *fig. 5*, which supplies water to various other parts of the plumbing system to be described later.

The only other tapping likely to be taken off the rising main before it reaches the loft is a branch pipe for an outside tap.

There are, of course, alternatives to having the cold water storage cistern situated in the loft. Sometimes a cistern is installed in the bathroom or in a cupboard on the landing.

While this sort of arrangement will mean that the cistern is less likely to be affected by frost, it may also mean, for one thing, that the w.c. cistern which it supplies will refill very slowly. For another thing, it may make it impossible to work a shower properly owing to insufficient water pressure or head, as it is called. This problem I will deal with later.

SUPPLY PIPES

Water is supplied to the cold water

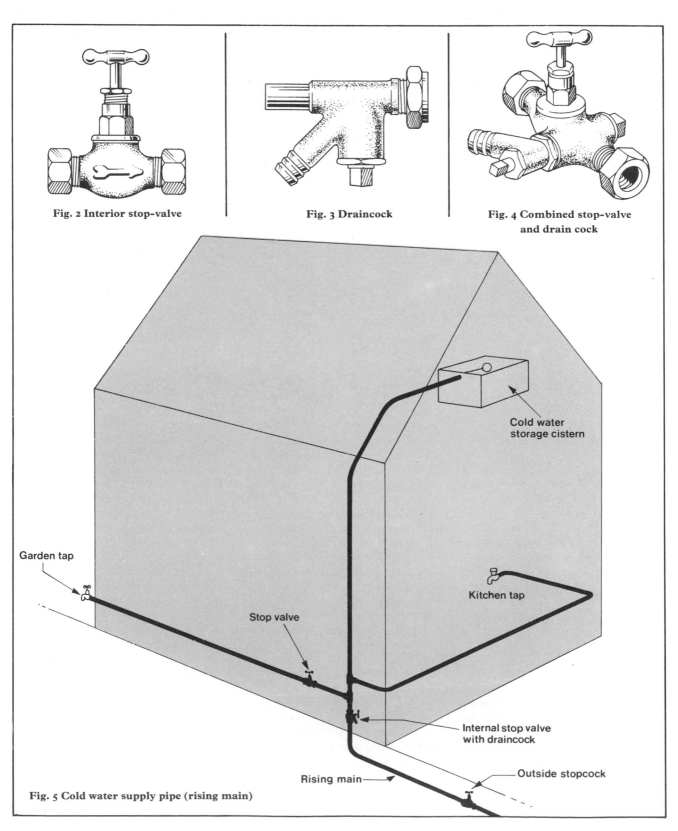

Fig. 2 Interior stop-valve

Fig. 3 Draincock

Fig. 4 Combined stop-valve and drain cock

Cold water storage cistern

Garden tap

Kitchen tap

Stop valve

Internal stop valve with draincock

Rising main

Outside stopcock

Fig. 5 Cold water supply pipe (rising main)

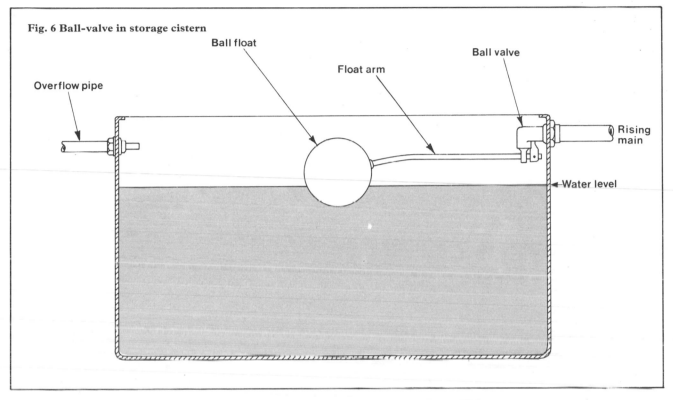

Fig. 6 Ball-valve in storage cistern

Ball float

Float arm

Ball valve

Overflow pipe

Rising main

Water level

storage cistern through a ball valve fitted in the cistern, at the end of the rising main, *fig. 6*. The cistern supplies cold water to the bath, the bathroom basin and the toilet cistern. It also provides a cold water supply to the hot water cylinder or tank, *fig. 7*.

You will find at least a couple of supply pipes leading from points about 2 in from the base of the cold water storage cistern. These are normally $\frac{3}{4}$ in in diameter. One leads to the hot water cylinder and the other supplies the bath and bathroom basin taps and w.c. cistern.

Each of these pipes should be fitted with a gateway valve, *fig. 8*. These valves enable supplies to the cold water system and to the hot water system to be turned off independently. This is a great advantage when repairs are necessary as only the circuit affected needs to be isolated from the water supply.

It is a sound idea to make sure both these stop-valves – and the one on the rising main – turn easily. Give them a

couple of turns occasionally and apply a little grease at the same time.

DIRECT AND INDIRECT

To recap. The rising main feeds the tap over the kitchen sink, possibly a tap in the garden and the cold water storage cistern. All other points are normally supplied from that cistern. This is an *indirect* cold water system.

In some older houses, a *direct* cold water system may be found. In this case the rising main feeds not only the tap over the kitchen sink, but branches are also taken from it to the bath, bathroom basin and w.c. cistern.

The direct method is not usually found in modern installations. In some older houses both systems are sometimes used. For example, one w.c. may be supplied direct from the rising main, while the second w.c. may be connected to a cold water storage cistern.

There is a way to find out if the system is direct or indirect. Turn off the stop-valve on the rising main. If this stops water flowing to all the cold

water taps and w.c. cisterns, the system is direct. If only the flow to the kitchen tap stops, you have an indirect system.

TYPES OF CISTERN

The most unusual type of cold water cistern is a galvanised steel one, at least in older homes. These cisterns are liable to rust unless their interiors are protected by a special type of paint or by other means which are explained in the chapter on corrosion.

Fortunately, there are other types of cistern available. Undoubtedly the least troublesome are the polythene and glass fibre types.

These, of course, cannot corrode or be damaged by frost and they are extremely tough though light in weight. Some polythene cisterns are even flexible enough to be bent and passed through small trap doors into the loft.

Some cisterns are made of asbestos cement but, although excellent in themselves, these are somewhat heavy and can be damaged more easily, either by accident or by frost.

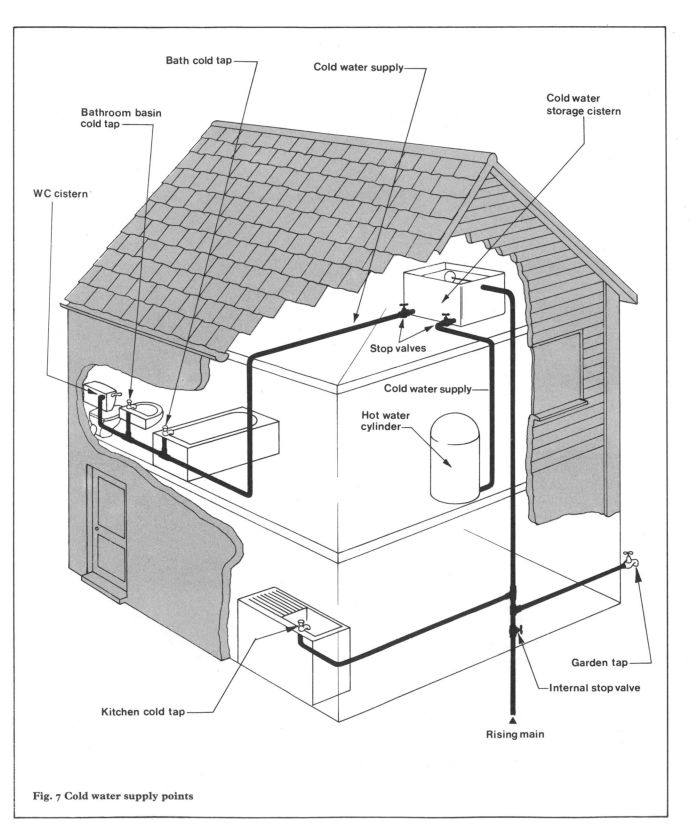

Fig. 7 Cold water supply points

Fig. 8 Gateway valve

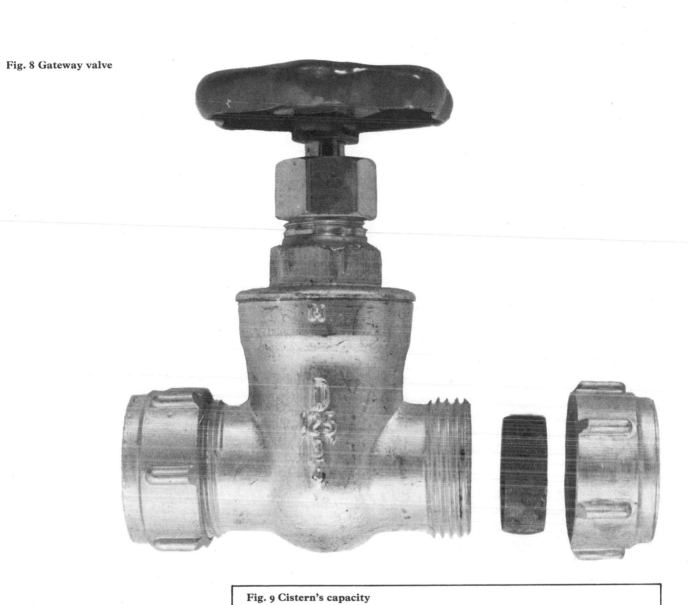

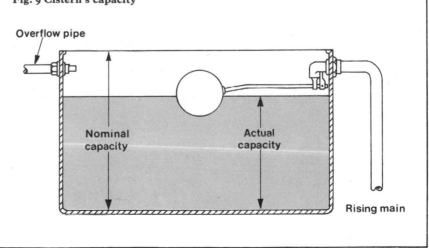

Fig. 9 Cistern's capacity

Overflow pipe

Nominal capacity

Actual capacity

Rising main

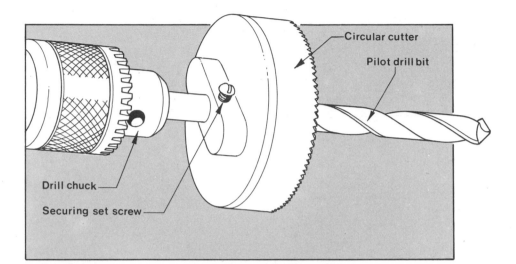

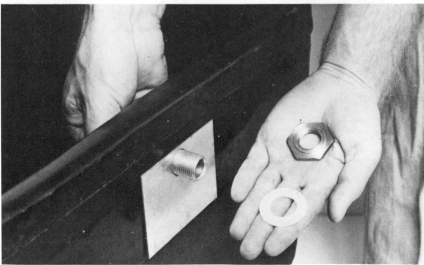

Fig. 10 Tank cutter

Fig. 11 Offering the stem of the float valve to the tank hole

Fig. 12 Offering the nut and plastic washer to the float valve stem

Fig. 13 Securing the nut to the stem

Fig. 14 Offering the tank connector to the float stem

Fig. 15 The tank connector fitted

Fig. 16 Copper rising main connected

(Photographs by courtesy of and Copyright Summerhill Heating Services Ltd)

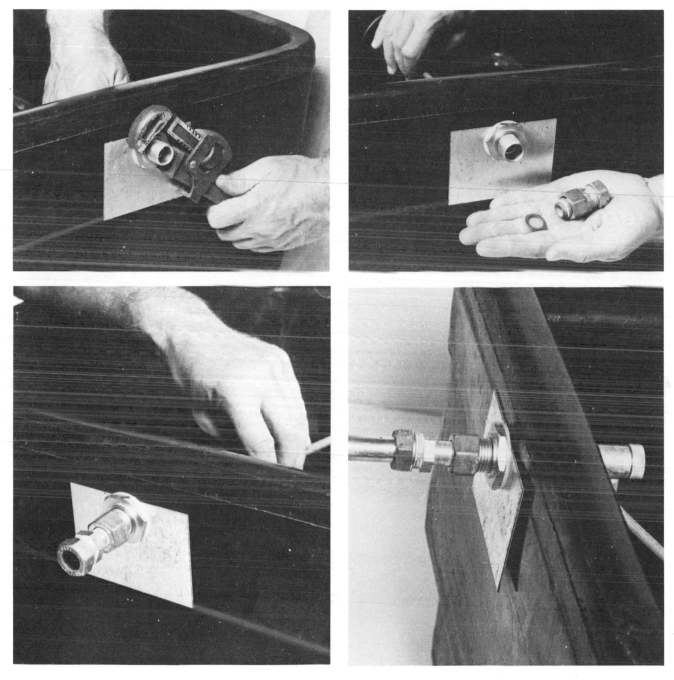

CAPACITIES

Most probably your cistern will be a galvanised type and its *actual* capacity will be 50 gallons. This is the normal maximum size stipulated by water boards in some parts of the country for a cistern which supplies hot and cold water systems. In other areas an 80 gallon capacity is the maximum requirement.

The quantities referred to represent the *actual* capacity of a cistern – to the point where the overflow pipe is filled, or rather just below it, *fig. 9*. If you see a cistern advertised as having a *nominal* capacity of, say, 60 gallons, this would be the amount it could carry if filled to its brim. Its *actual* capacity would be 50 gallons. A cistern with a *nominal* capacity of 100 gallons will have an *actual* capacity of 84 gallons.

These figures will be of interest to you if it becomes necessary to buy a

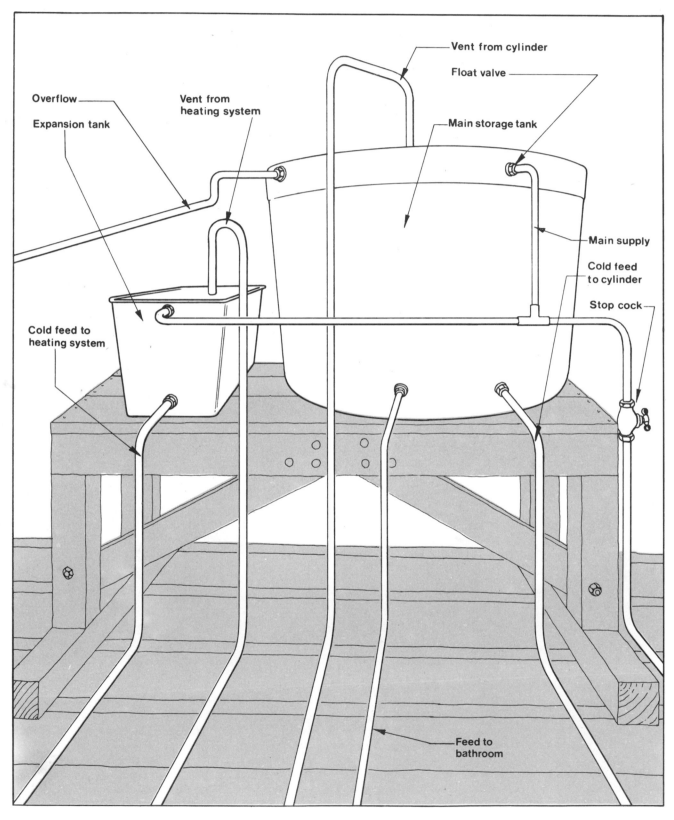

Fig. 17 The installation of a cistern on a firm platform base, showing pipe connections for feed and supply

new one. You can get a good idea of an existing cistern's capacity by measuring it.

A galvanised cistern measuring 30 inches long by 24 inches wide by 24 inches deep will have a nominal capacity of 60 gallons (50 gallons actual). An asbestos cement cistern of similar capacity will be about one inch larger in each dimension.

INSTALLING A CISTERN

An advantage of the round polythene cisterns is that they are fairly simple for a d-i-y man to install. It is not a difficult job to cut holes in a cistern of this type for the supply pipe connections.

The recommended distance from the base of a cistern for the supply pipe hole is 2 inches. The reason for leaving a space is to prevent grit and other foreign matter washed in from the main finding its way into the pipes.

Obviously, the greater the distance is, the more the capacity of the cistern will be reduced, but there will be less risk of debris entering the pipes.

In most cases it will be easier to cut the holes and fit the connectors *before* the cistern is taken into the loft.

BALL-VALVE HOLE

Near the top of the cistern a hole has to be made for the ball-valve. In most cases the centre of the hole will be $1\frac{1}{2}$ in from the top of the cistern, but its position may be dictated by a backplate which is fitted by some makers to support the ball-valve's weight.

This plate has a flange positioned against the cistern's flange which runs around its top. The tapping for the ball-valve is indicated by the position of the hole in the backplate.

The holes can be cut with a hole saw or a centre bit hole cutter or, better still, a hole saw attachment fixed in an electric drill.

To fix the pipes and ball-valve to the cistern, flange nuts are used.

A polythene or rubber washer should be fitted between the cistern and flange nut on the outside of the cistern.

FIRM BASE

It is essential that polythene cisterns should stand on a firm and flat platform, *fig. 17*. The joists in the loft are not by themselves sufficient support.

Before attempting to replace a cistern, check that all the pipes involved are long enough to reach the holes in the new cistern. If not, they can be lengthened or shortened as described in the chapter on Pipes and Joints.

This is a point worth watching if you are planning to install a shower. In this case you may have to raise the level of the cistern to provide greater water pressure to the shower. And if you do this, make certain the cistern can be safely supported. Although these modern cisterns are light to handle, one gallon of water weighs 10 lb! The weight of the whole thing is therefore considerable, so a very strong timber framework must be made.

OVERFLOW PIPE

The overflow pipe position is very important. Its distance from the top of the cistern must be at least twice the pipe's diameter. Also, the bore of the pipe must be larger than that of the inlet – at least $\frac{3}{4}$ in.

Note, too, that the height of the overflow pipe above the water must be at least one inch, or not less than the internal diameter of the pipe – whichever is the greater.

Make sure the overflow pipe has a slight fall to prevent water lying in it and possibly freezing.

The cistern should be covered with a lid to prevent insects entering and debris falling into it. Some polythene cisterns are supplied complete with lids.

Although the measurements quoted here are in imperial sizes, the plumbing industry has now 'gone metric'. We no longer talk of $\frac{1}{2}$ in pipe but 15 mm. The apparent discrepancy in the two measurements is explained on page 49, Pipes and Joints.

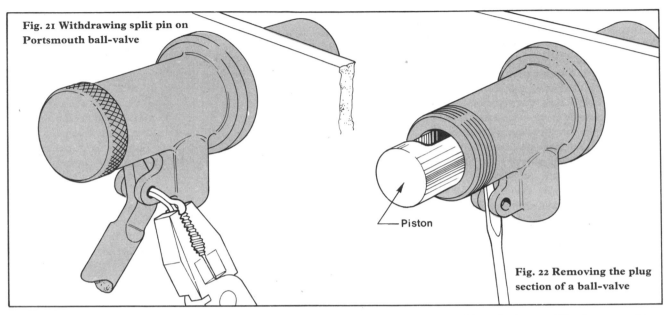

Fig. 21 Withdrawing split pin on Portsmouth ball-valve

Piston

Fig. 22 Removing the plug section of a ball-valve

cure the trouble temporarily, but for a more permanent solution the ball-valve should be dismantled, cleaned and greased with Vaseline.

A steady drip from the overflow pipe is a sign that the ball-valve needs re-washering. As this involves dismantling, we can treat this job and that of cleaning the ball-valve as a combined operation.

REWASHERING

To rewasher a Portsmouth type ball-valve, you will need a ½ in ball-valve washer. Turn off the water at the stopcock or stop-valve as described. Unscrew the cap (if there is one) on the body of the valve. Now close the ends of the split pin on which the float arm pivots. Do this with pliers – carefully or the pin may snap if it is corroded. Then withdraw the split pin, *fig. 21*.

Put a screwdriver blade in the slot on the valve body and push the plug section of the valve out of the open end, *fig. 22*. Make sure you don't drop it in the cistern!

The plug is in two parts, one of which is the washer retaining cap. This should be unscrewed with pliers, but be careful not to damage the valve body.

If you find the cap difficult to un-screw, an alternative is to pick out the old washer with a penknife. The new washer can then be persuaded to go under the lip of the cap, but make sure it fits firmly on the seating.

While the valve is dismantled, clean the plug inside and out with a fine grade of emery paper, grease it with Vaseline and reassemble in the body of the valve.

Croydon ball-valves are rewashered in a similar way to Portsmouth types, but if you have a Croydon valve fitted and it gives trouble, it is well worth considering replacing it with a modern diaphragm type.

DIAPHRAGM BALL-VALVES

If, after replacing the float arm, the overflow trouble persists, the seating of the ball-valve may be damaged. To reseat the valve you need a special reseating tool.

The expense of this tool and the effort involved is hardly worth while. It is better to get a new ball-valve, preferably a diaphragm pattern, *fig. 23*, which is less likely to leak or jam than a Portsmouth or Croydon type.

A diaphragm valve has a replaceable nylon nozzle which is closed by a

rubber diaphragm. This is operated by a plunger which passes through the backplate of the valve. The plunger is pressed against the diaphragm by the float arm as it pivots upwards when the water rises in the cistern.

Diaphragm valves rarely go wrong, but occasionally the diaphragm may jam against the nozzle. This will reduce the flow of water to a mere trickle.

Fortunately, these valves are simple to take apart. If the valve jams, un-screw by hand the knurled cap, *fig. 24*, and pick the diaphragm from the nozzle with a penknife. After cleaning and reassembling, the valve should work normally again.

WATER PRESSURES

Ball-valves are designed for low, medium or high water pressures. This is indicated by the diameter of the nozzle orifice.

If a low pressure valve is fitted to a pipe which is under mains pressure, this can cause a leak from the overflow pipe. The majority of ball-valves fitted to cold water storage cisterns are under high pressure and it is unlikely that the low pressure valve will have been fitted in error.

Ball-valves fitted to w.c. cisterns may be under low *or* high pressure. Most

w.c.s, however, will be supplied by the
storage cistern and will therefore be
under low pressure.

If a high pressure valve were used
in such an installation the w.c. cistern
would fill very slowly. In persistant
cases of slow filling a full-way valve
may have to be fitted.

As a rule, ball-valves are stamped
HP or LP (high pressure or low
pressure). Diameters of nozzles
are: high pressure, $\frac{1}{8}$ in; medium
pressure $\frac{3}{16}$ in; low pressure, $\frac{1}{4}$ in;
full-way, $\frac{3}{8}$ in or $\frac{1}{2}$ in.

NOISY VALVES

One of the greatest annoyances some
ball-valves cause is noise when they
are operating. Much of this is caused
by ripples forming on the surface of
the water as water flows in to the
cistern. This causes the ball float to
vibrate and the rest of the plumbing
system involved magnifies the noise
and at times causes water hammer
(hammering in the pipes).

Many existing ball-valves are fitted
with plastic silencer tubes screwed to
the valve outlet, *fig. 25*. Through this
tube the inflowing water discharges
below the water surface – thus reducing
the ripples.

Unfortunately, from the point of
view of noise, anyway, these tubes have
been banned by the British Water-
works Association. The reason for this
is the possible danger of contaminated
water from a storage cistern being
siphoned back to contaminate the main
water supply.

At the time of writing, there has
been no move to ban existing silencer
tubes, but they should not be fitted in
new installations.

OVERHEAD OUTLETS

Manufacturers have produced alternative ways of silencing the incoming
rush of water and no doubt will continue to improve upon these methods.

Some diaphragm ball-valves are now
designed with an overhead outlet,
fig. 26. In one case, at least, this
outlet can be adjusted so that the spray

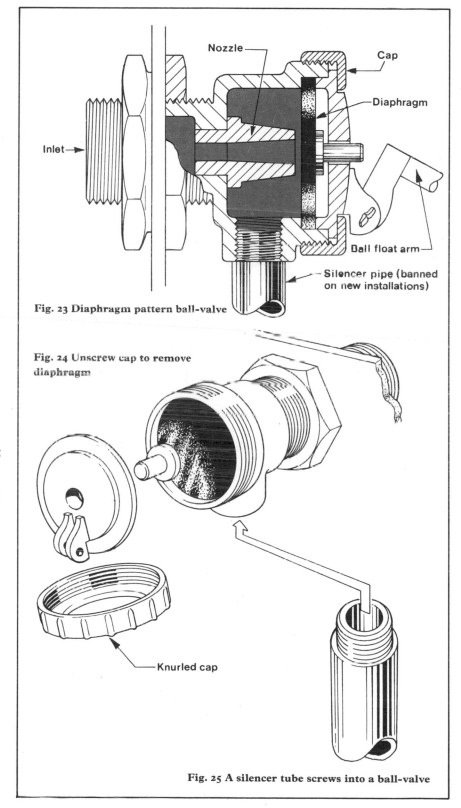

Fig. 23 Diaphragm pattern ball-valve

Fig. 24 Unscrew cap to remove diaphragm

Fig. 25 A silencer tube screws into a ball-valve

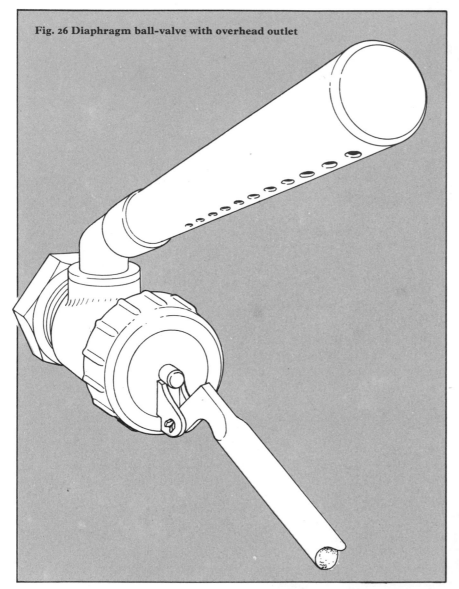

Fig. 26 Diaphragm ball-valve with overhead outlet

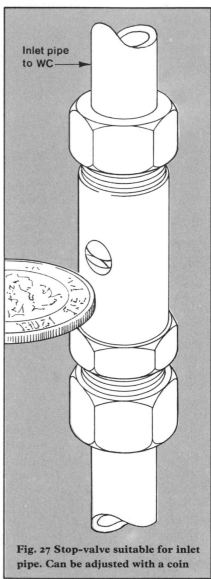

Inlet pipe to WC

Fig. 27 Stop-valve suitable for inlet pipe. Can be adjusted with a coin

of incoming water hits the side of the cistern instead of the water surface.

There are other ways in which a noisy water supply can be quietened. The pipe attached to the ball-valve, for instance, should be firmly anchored to a permanent support such as the roof timbers.

EQUILIBRIUM VALVES
If your cistern suffers from severe water hammer, or if the water pressure is variable, it may be worth while fitting an equilibrium valve. Although more costly than ordinary ball-valves, these types have a smooth, quiet action.

An equilibrium valve has a channel drilled in its plug which lets water pass through to another water chamber behind it. So there is equal pressure behind and in front of the plug. In other ball-valves, water pressure is constantly trying to force the valve open from one side only.

When a ball-valve on a w.c. cistern needs attention, it usually means that the bath and the bathroom basin taps are also deprived of water when it is turned off at the stop-valve. To avoid this, it is worth fitting a simple stop-valve in the pipe to the w.c. ball-valve, fig. 27. (A suitable valve is the Markfram control valve. Another is the Nevastop.)

This can be fixed anywhere along this pipe, but preferably near the w.c. cistern. It needs only a screwdriver or coin to turn it off or on and can be fitted easily by using compression joints. These are dealt with under Pipes and Joints, page 41.

Hot Water Systems

Once the layout of the cold water supply system is understood, you will find it easier to appreciate the workings of the hot water system. Although the hot and cold water systems work partly in conjunction with each other, I have omitted most of the cold water pipes in *figs. 28* and *29* so that the hot water system can be more readily understood. In *fig. 30*, however, I have combined both hot and cold systems in one diagram, illustrating an indirect hot water system.

I must emphasise that the arrangements shown in these diagrams are *basic* layouts. Variations may be found in some systems, especially in older houses where the plumbing has been extended or altered from time to time.

In older houses some pipes may appear to serve no purpose whatever! It may be that they form part of an old plumbing system now bypassed by a later design.

The safest course to follow in these circumstances is to leave well alone until you can get advice from an expert. Plumbing in some ways is rather like electricity. It's perfectly safe to carry out alterations and repairs yourself provided you know exactly what is involved!

TWO SYSTEMS

I will now describe a basic hot water system, which is not to be confused with a central heating system. For the moment I will deal solely with a supply of water for washing, bathing and similar purposes, heated by a boiler.

There are two types of hot water systems – direct and indirect.

A typical *direct* system comprises a hot water cylinder which is fed with water from the cold water storage cistern in the loft, *fig. 28*.

The water flows down what is called the return pipe from the *bottom* of the hot water cylinder to the *bottom* of the boiler. There the water is heated and rises up the flow pipe from the *top* of the boiler to the *top* of the cylinder.

As it enters the cylinder, the hot water pushes the cold water down the

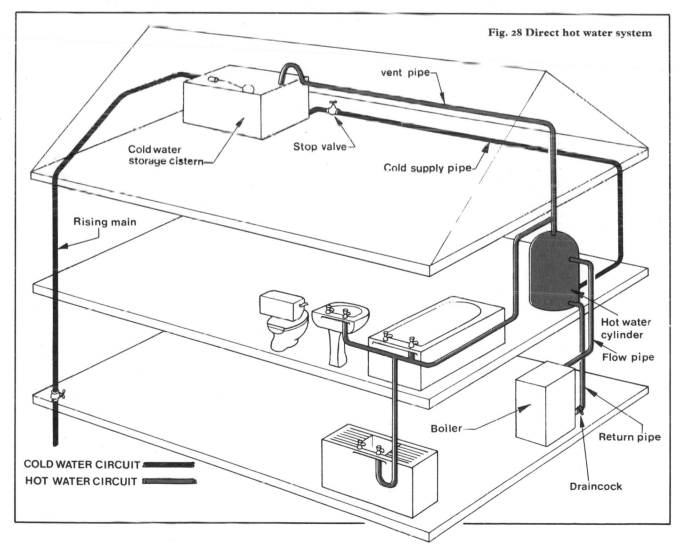

Fig. 28 Direct hot water system

vent pipe

Cold water storage cistern

Stop valve

Cold supply pipe

Rising main

Hot water cylinder

Flow pipe

Boiler

Return pipe

Draincock

COLD WATER CIRCUIT
HOT WATER CIRCUIT

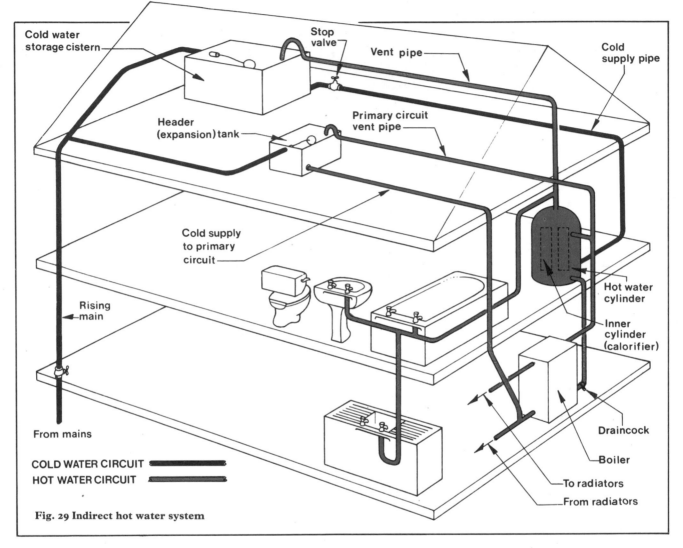

Cold water storage cistern

Stop valve

Vent pipe

Cold supply pipe

Header (expansion) tank

Primary circuit vent pipe

Cold supply to primary circuit

Rising main

Hot water cylinder

Inner cylinder (calorifier)

From mains

Draincock

Boiler

To radiators

From radiators

COLD WATER CIRCUIT

HOT WATER CIRCUIT

Fig. 29 Indirect hot water system

return pipe to the boiler to be heated.

This circulation or cycle continues until all the water in the boiler and cylinder circuit is hot.

BOILING POINT

If the boiler were allowed to burn continuously without any water being run off from the hot water taps, the water would boil. You may have noticed your boiler makes a noise rather like a boiling kettle when the water is very hot. This is a sign that the water is near, or has reached, boiling point.

Boilers should never be allowed to boil. The water can usually be cooled down sufficiently by running some off from the hot water taps. However,

there is a safety device built into hot water systems which safeguards them against boiling. This is the vent pipe.

This pipe runs from the top of the cylinder up to the cold water storage cistern. The pipe is open at its top end and is normally bent over the cistern. If the water in the cylinder boils, it is pushed up the vent pipe to discharge into the cold water storage cistern.

Supplies of heated water to the hot water taps come from a branch pipe which is connected, just above the hot water cylinder, to the vent pipe.

A draincock should be fitted at the lowest point of the system. It is usually on the return pipe near the boiler. The draincock enables the whole system to

be emptied if necessary.

The direct method of heating water is a very common one and the principles involved in it are similar whether the hot water is stored in a cylinder or in a rectangular galvanised steel tank. And the same principles apply to any method of heating – solid fuel, gas, oil or electricity.

INDIRECT SYSTEM

Two separate circuits are used in an *indirect* system, *fig. 29*. One, called the primary, is from the boiler to a small cylinder, or calorifier, fitted in the main hot water cylinder. Water cannot be drawn off the primary circuit (unless the system is drained off). Therefore

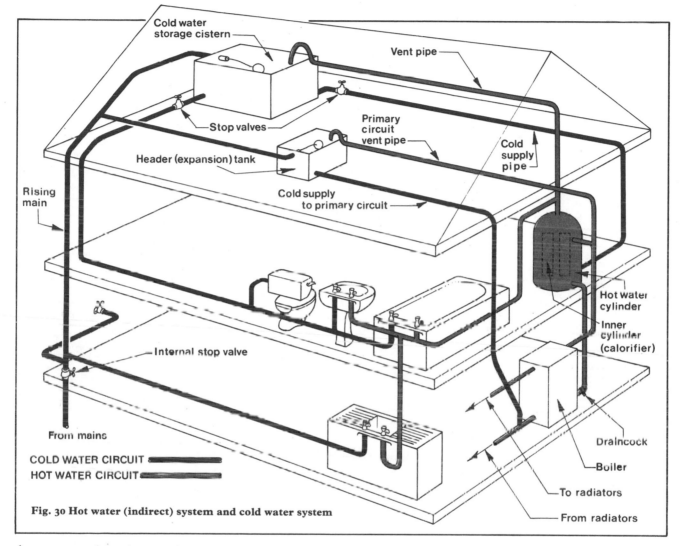

Fig. 30 Hot water (indirect) system and cold water system

the same water is used over and over again.

Water in the *main* cylinder is heated indirectly by the inner cylinder and is supplied to the hot taps.

The inner cylinder normally gets its own supply of cold water from a small header or expansion tank. This is needed only to replace losses caused by evaporation or for renewing the water supply to the inner cylinder if the whole system is drained off.

The header tank's capacity is usually around ten gallons and it has its own vent pipe. Some types of indirect cylinder, however, do not need a header tank.

If radiators are added to an indirect system they are supplied by the primary circuit.

It is essential in most cases that an indirect hot water system should be used if any form of central heating is installed in the system.

How can you tell which system is installed in your house? Look at the connections on the cylinder. On an indirect cylinder both the primary connections are male. *Figures 31* and *32* show two types of indirect cylinder to illustrate this point.

The secondary and cold water connections are female. On a direct cylinder all the connections are female, *fig. 33.*

ADVANTAGE
A big advantage of an indirect system is comparative freedom from scale in hard water areas or from corrosion in soft water areas. (Scale and corrosion are dealt with in a separate chapter).

HEADER TANK
One purpose of a header tank in an indirect system is to allow for expansion of the primary circuit when it has been heated. Expansion can cause the volume of water in the circuit to be increased by several gallons.

Therefore, the ball-valve in the header tank should be adjusted so that it cuts off the supply when only 2 or 3 inches of water are in the tank. When expansion takes place, the water will rise above the level of the ball float but should not rise up to the level of the ball-valve.

Incidentally, as this ball-valve will not need to operate very often, to replace water lost through evaporation, it may jam when it is in either the open or closed position. A careful eye should therefore be kept on it.

To be on the safe side, it is better to change the ball-valve, if it is a Portsmouth or Croydon type, to a modern diaphragm valve which was described earlier in the chapter on ball-valves.

AIR LOCKS

One of the most common problems in hot water systems is an air lock. Sometimes you may find that hot water will flow from a tap evenly at first, but soon it will be reduced to a mere trickle, or the water may stop flowing entirely.

There can be several reasons for air locks forming, some of which I will outline. First, though, the cure.

Fit one end of a length of garden hose to the tap which has the air lock and the other end to the cold water tap in the kitchen, *fig. 34*. When both taps are fully turned on, the pressure from the main should force the air bubbles out of the pipe and up through the vent pipe.

MAINS PRESSURE NEEDED

The kitchen tap is the one to use because normally it will be the only one which is under mains pressure. The pressure at the taps supplied by the cold water storage cistern will not usually be great enough to shift the air bubble.

If an air lock forms in a pipe supplying a bath tap, it is unlikely that your hose pipe will fit the tap. The solution here is to connect the hose to the bathroom basin tap instead. Both taps are fed by the same supply pipe so the air lock can be dispersed from either.

One possible cause of an air lock is that the diameter of the pipe supplying cold water to the hot water cylinder, *fig. 35*, is too small. It should be at least $\frac{3}{4}$ in in diameter. A $\frac{1}{2}$ in pipe fitted here in error can cause air locks.

Other possible causes of air locks include:

1. A $\frac{1}{2}$ in diameter stop-valve fitted in error on the cold water supply pipe to the hot water cylinder. The remedy for this is to change the stop-valve to a $\frac{3}{4}$ in type.

2. The stop-valve is partly closed. It should, of course, be kept fully open or its diameter will, in effect, be reduced.

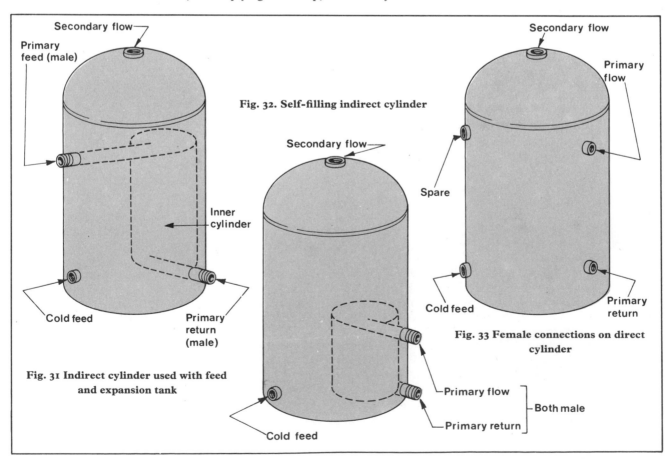

Fig. 31 Indirect cylinder used with feed and expansion tank

Fig. 32. Self-filling indirect cylinder

Fig. 33 Female connections on direct cylinder

3. The cold water storage cistern is too small to supply enough water for both the bath taps. The water level in the cistern may drop to below the draw-off point before the bath is filled.

Check that the *actual* capacity of the cold water cistern is at least 50 gallons.

4. Check that the draw-off point for the hot water cylinder is no more than 2 in from the base of the cold water storage cistern (X in *fig. 35*). If the distance is greater than 2 in, the capacity of the cistern will be reduced accordingly.

5. Hot water supply pipes are exactly horizontal or rise slightly.

In a correctly designed system, these pipes should, of course, *fall* slightly, *fig. 36*, from the point at which they are connected to the vent pipe, to the point where they rise to be connected to the hot water taps. The vent pipe is there to let any trapped air escape, so any air bubbles which form in the pipes should be able to reach the vent pipe easily.

6. If the hot water system has to be drained entirely for any reason, the whole of the hot water circuit will be filled with air.

REFILLING

To prevent air locks forming when refilling the system, connect a hosepipe to the draincock beside the boiler and to the cold tap in the kitchen. Turn both the taps on and the cistern will fill upwards, the water driving the air out of the system in front of it.

There are other causes of an inadequate flow from hot water taps which can be due to a badly designed system. For example, if the boiler is not standing absolutely level an air bubble can form in the boiler itself. This can cause loud banging sounds in the system.

Trouble can also arise if the flow and return connections between boiler and cylinder are not correctly made. The correct method of making these connections is shown in *fig. 30*.

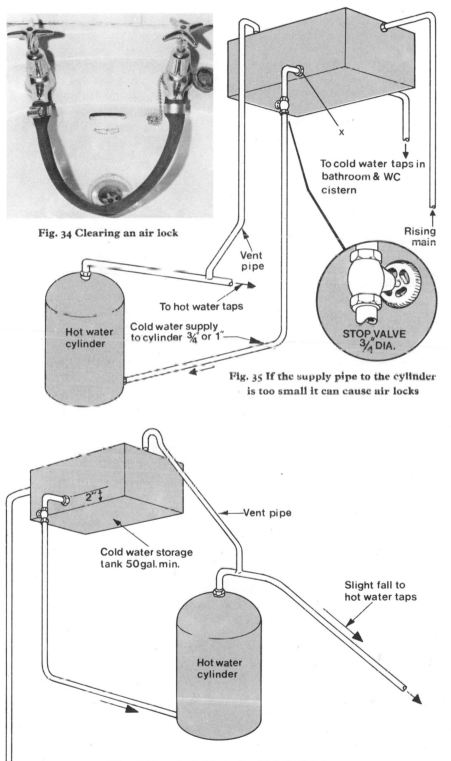

Fig. 34 Clearing an air lock

To hot water taps

Vent pipe

To cold water taps in bathroom & WC cistern

Rising main

Cold water supply to cylinder ¾" or 1"

Hot water cylinder

STOP VALVE ¾" DIA.

Fig. 35 If the supply pipe to the cylinder is too small it can cause air locks

Vent pipe

2"

Cold water storage tank 50 gal. min.

Slight fall to hot water taps

Hot water cylinder

Fig. 36 Pipes to hot taps should fall slightly

Toilet Cistern Troubles

One room in the house which is often more likely to develop plumbing troubles than any other room is the lavatory. And the piece of equipment frequently responsible is the flushing cistern.

It may be noisy, its ball-valve may jam or it may refuse to flush the first time it is pulled. These are the most common problems which can arise.

First a brief outline of how flushing cisterns work. They are normally supplied with water from the cold water storage cistern in the loft, but some are supplied direct from the main.

BALL-VALVE INLET
In each case, the water enters the cistern through a ball-valve inlet as in the cold water storage cistern. Occasionally the toilet cistern may overflow due to the ball-valve jamming or its washer becoming worn out. If that happens, the treatment is the same as that described in Ball-valves, page 17.

Turn off the water from the internal stop-valve. If there isn't one, you will have to drain off the main storage cistern after turning off the water from the external stopcock. Flush the toilet cistern to empty it, then proceed as described on page 17.

While you are at it, clean out any rust flakes or other debris from the base of the flushing cistern.

FLUSHING PROBLEMS
A most annoying and embarrassing situation is a flushing cistern which won't flush the first time. Equally irritating is a cistern which continues to discharge after the flushing process is completed.

Let's take first the old-fashioned high level type, the Burlington pattern flushing cistern, *fig. 37*. Its base contains a well in which stands a heavy bell. The bell is joined to a lever which pivots on a support fixed in the cistern. The chain is attached to the other end of the lever.

Pull the chain and the bell is raised. Release the chain and the bell falls. Its weight traps water in the well, forcing the water level in the bell to rise, and some water flows over a stand pipe into the flush pipe.

The stand or siphon pipe is connected to the flush pipe and rises under the bell to a point above the normal water level.

VACUUM
When this water passes through the flush pipe it creates a partial vacuum. The air pushes water in the cistern under the bell down through the flush pipe and flushes the toilet pan.

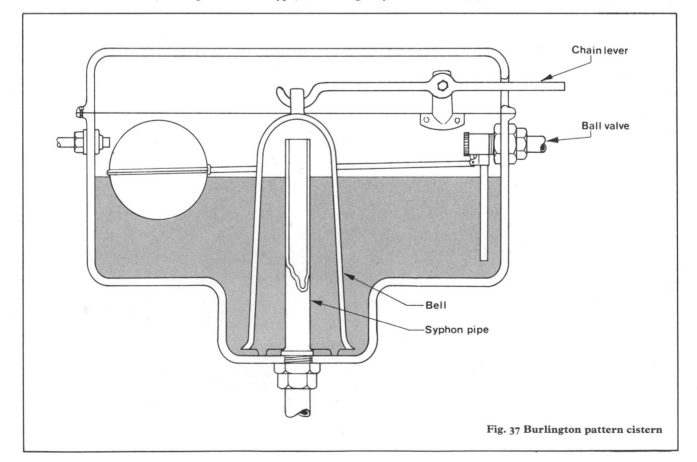

Chain lever

Ball valve

Bell

Syphon pipe

Fig. 37 Burlington pattern cistern

The flow of water continues until air gets under the bell and breaks what is called siphonic action.

It will be seen that this is not a complicated cistern, but it can be a very noisy one. Not a great deal can go wrong with a Burlington, but after a time grit and other rubbish may collect in the well of the cistern. This causes water to continue to flow down the flush pipe after the flushing operation is completed.

The cure, of course, is to clean the cistern out – if you can get at its bottom, that is! Access can usually be obtained by lifting off the lid.

Occasionally this type of cistern will give trouble after a new ball-valve has been fitted. The flow of water into the cistern is then usually so rapid compared with when the old valve was operated, that air is not able to reach the rim of the bell. The result is again continuous siphonage.

To cure this, it should be sufficient partly to close the stop-valve on the inlet pipe. If no stop-valve is fitted, this should be done.

CORROSION

Burlington pattern cisterns are also subject to corrosion inside and out. They can be wire brushed to remove loose rust and treated with a chemical rust remover to get rid of the rest.

The interior of the cistern can then be painted, when dry, with bituminous paint and the exterior with ordinary paint.

However, if the cistern gets into a state such as I have described, you may feel that all the repair and maintenance necessary is hardly worth the effort, and you will probably consider having the cistern replaced with a modern direct action type

If it is desired to keep a new cistern at a high level, you can get one which has a well bottom. These are specially designed as replacement cisterns for the Burlington models.

A far better solution, though, is to have a modern, low-level suite fitted. This type is much easier to get at when things go wrong.

LOW-LEVEL CISTERN

Now let's see what faults can arise with low-level direct action cisterns. First, a look at how these cisterns work, however.

Most of them have a flat base and a siphon pipe which is connected direct to the flush pipe. The siphon pipe rises above the level of the water. It is joined to an open-ended dome below the water which terminates just above the base of the cistern. The design may vary according to the model.

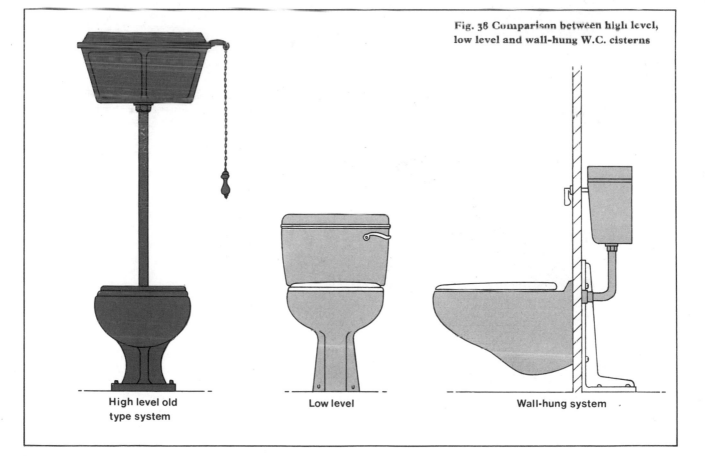

Fig. 38 Comparison between high level, low level and wall-hung W.C. cisterns

High level old type system

Low level

Wall-hung system

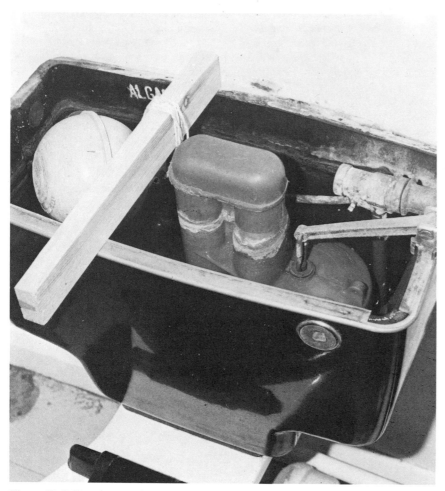

Fig. 39 Ball float arm tied up

When the lever is pulled to flush the toilet, a disc with one or more holes in it rises in the dome. A diaphragm or flap valve covers these holes and does not allow any water to pass the disc. This action causes water to be thrown over the bend in the siphon pipe into the flush pipe.

As the water falls, it mixes with air and makes a partial vacuum. The water pressure now increases behind the flap valve and forces it away from the disc. Water flows through the holes and the cistern empties.

THE FLAP VALVE

It is failure of the flap valve which is usually to blame when the cistern refuses to flush at the first pull or even after several attempts.

Renewing the flap valve is not a difficult job, but it is not really one I would recommend for anyone who has no plumbing experience. Once you have seen the job done, however, you will probably say, 'nothing to it'.

New flap valves can be obtained from plumbers' or builders' merchants. They are made in various sizes, but you can safely choose the largest size. As they are made of thin plastic they can be easily trimmed to size with scissors if too large. It is a good plan to keep one or two spares handy once you know the size required.

To replace a faulty flap valve, first turn off the water at the internal stop-valve or tie up the ball float arm, *fig. 39.* This will prevent more water entering the cistern. Flush the cistern to empty it of water.

The next job is to unscrew the nut which holds the flush pipe in position, *fig. 40.* Have a bowl or bucket handy to catch what water there is left in the cistern after flushing.

Immediately above this nut is a larger one just below the cistern. Unscrew this next and use a wrench with jaws that open wide. After removing this nut you can then lift out the siphoning mechanism complete.

Now disconnect the rod or hook which is joined to the operating lever,

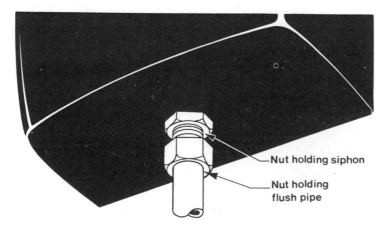

Fig. 40 Remove these nuts to release siphoning mechanism

Nut holding siphon

Nut holding flush pipe

fig. 41. Pull up the disc and the rod which operates it and you will see the old flap valve, or what is left of it.

TRIM TO SIZE

Remove this and replace it with your new plastic flap valve and trim it to the same size as the disc it is to cover. The valve must not foul the sides of the siphon.

(Means of access to the disc and flap valve will vary according to the make of cistern. Some siphons are in two parts which are bolted together near the base. Make sure you know the order in which to put back all the bits and pieces.)

Before reassembling all the components in reverse order, clean out the cistern and the siphon. All sorts of rubbish tends to collect in the bottom of a w.c. cistern and if any of it gets under the siphon, it can cause water to trickle continuously down the flush pipe after the cistern has been flushed. In other words, continuous siphonage takes place, as already described.

WATER LEVEL

Before you put the lid back on the cistern, restore the water supply, flush the pan and note the level to which the water has risen in the cistern when the ball-valve closes and water stops flowing in.

This level should be about $\frac{1}{2}$ in below the overflow pipe outlet. If it is lower than this, you may find that when the cistern is flushed the pan is not completely cleared.

This is an irritating fault sometimes caused because the flush is inadequate. Also, if the water level is too low, the cistern may fail to flush at the first pull.

You can control the water level either by bending the float arm (if a metal type) up or down or by adjusting the screw on the arm as described earlier.

Another reason for an inadequate flush can be that something is obstructing the flush pipe at the point where it is connected to the toilet pan. Or there may be an obstruction around the rim of the pan. The only way you can check this is by holding a small hand mirror in the pan under the rim.

When a flushing cistern is working properly, the water should run around each side of the pan in two cascades and meet in the centre of the front.

Sometimes a toilet cistern is suspected of having sprung a leak when its exterior surface is covered in moisture which, in severe cases, may drip to the floor.

Usually this is due to condensation and the answer to the problem is to increase ventilation in the room and to introduce some form of dry heat – a heated towel rail, for example, or a radiant heater fixed high up on the wall.

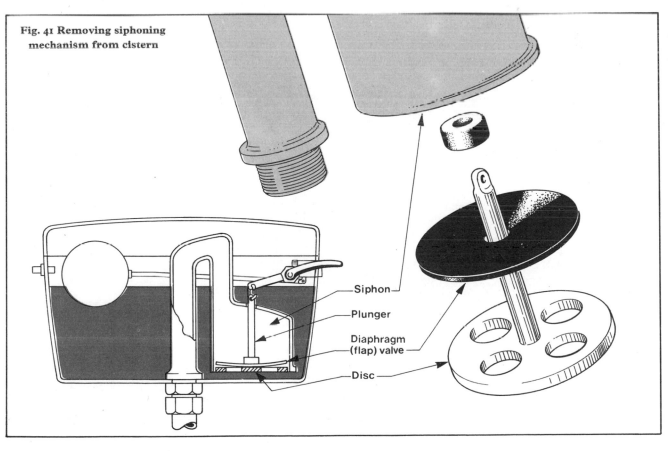

Fig. 41 Removing siphoning mechanism from cistern

Siphon

Plunger

Diaphragm (flap) valve

Disc

Taps Old and New

Rewashering a tap is a job every house-holder should be able to do. It is a simple operation, especially if the taps are modern types.

There is a wide variety of taps, but generally speaking most of them are based on two patterns. One is the bib tap, *fig. 48*; the other is the pillar tap, *fig. 49*.

Bib taps used to be the type fitted at kitchen sinks and are still fitted in some cases. These have a horizontal inlet. Pillar taps have a vertical inlet and are more likely to be found in bathrooms and wash basins. The procedure for rewashering is similar in each case.

An indication that a tap needs a new washer is, of course, a steady drip which persists even when the tap is tightly turned off. But there may be other causes to account for the drip.

WASHER SIZES
If you suspect that the washer is faulty on a cold water tap, get a new washer before dismantling. For cold water taps on a kitchen sink or bathroom basin you normally need a $\frac{1}{2}$ in washer, but for bath taps the size is $\frac{3}{4}$ in (a garden tap is usually fitted with a $\frac{1}{2}$ in washer).

I suggest that once you know the sizes of the washers required, you keep a small supply by you, or better still, buy the washers complete with jumpers, *fig. 50*.

The reason I suggest this is because old washers are sometimes very difficult to remove from their jumpers. The majority of washers on sale now are suitable for both hot and cold water taps, but check this when you buy them.

For rewashering taps you need two wrenches – one for holding the body of the tap and the other for turning the headgear. One wrench should prefer-ably have slim jaws so that they can fit under the easyclean cover.

TURN OFF THE WATER
First turn off the water supply to the tap. If it is a mains tap, turn off the

There is a wide range of taps available for modern plumbing installations

Fig. 42 Bath mixer taps

Fig. 43 Mixer taps for a double sink unit

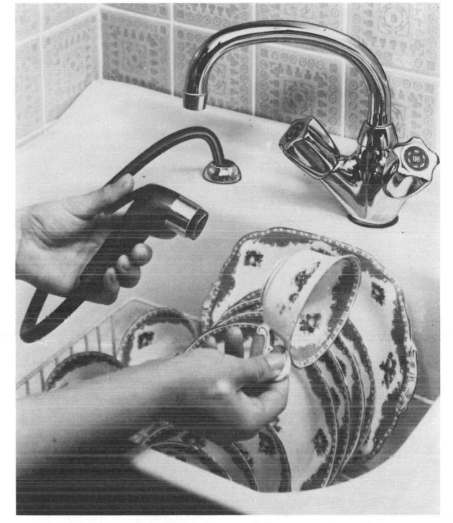

Fig. 44 Sink mixer with hot water spray attachment

Fig. 45 Wall mounted Bibcock

Fig. 46 Pillar tap for sink

Fig. 47 Wash basin tap

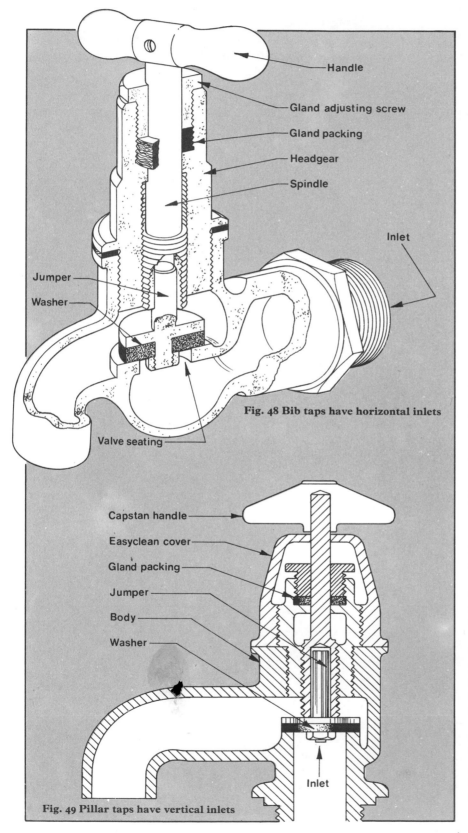

Handle

Gland adjusting screw

Gland packing

Headgear

Spindle

Inlet

Jumper

Washer

Valve seating

Fig. 48 Bib taps have horizontal inlets

Capstan handle

Easyclean cover

Gland packing

Jumper

Body

Washer

Inlet

Fig. 49 Pillar taps have vertical inlets

internal stop-valve on the rising main or the stopcock in the pavement as described earlier.

For the bathroom or other taps which are fed from the storage cistern, turn off the stop-valve on the supply pipe.

If there is no stop-valve on the supply pipe, you will have to drain all the water from the cold water storage cistern.

To do this, tie up the ball float arm so that the ball-valve closes and no more water can enter the cistern. Then open the bathroom cold taps and run off all the water in the cold water cistern.

If the tap to be rewashered is a hot water tap, you will still have to follow this procedure if there are no stop-valves on the supply pipes.

Having emptied the cold water cistern, turn on the hot water tap which is to be rewashered and let the water flow. You will probably find you will lose very little hot water.

The reason for this is that the hot water cylinder is emptied from its *top*. Once the water in it has fallen below the level of its discharge pipe, water will cease to flow.

Now turn on the tap as far as it will go. If there is an easyclean cover on the tap, this should be raised first.

STUBBORN COVERS

Here you will probably meet your first snag, especially if the cover hasn't been unscrewed for years. Sometimes you can unscrew the cover by hand, but stubborn covers will have to be shifted with a wrench.

If you have to use a wrench, protect the chromium plating of the tap by padding the jaws. In very stubborn cases it helps to apply a little oil between the easyclean cover and the tap spindle, *fig. 51.*

Another way to move an easyclean cover is to pour boiling water over it. Then try to shift it wearing a glove. As the metal expands with the heat, any scale deposits will be loosened and the cover should be easier to turn.

Once the cover is raised, *fig. 52a,* you

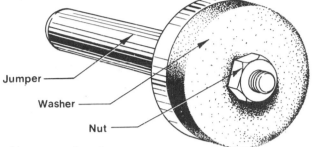

Jumper

Washer

Nut

Fig. 50 Combined jumper and washer

Fig. 51 Oiling a tap spindle helps to
remove a stubborn cover

Fig. 52a Unscrewing an easyclean cover

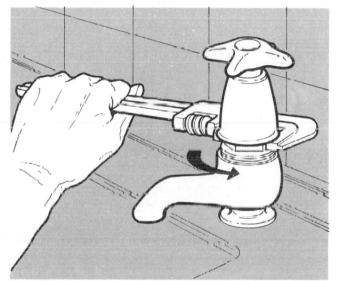

Fig. 52b Removing the headgear

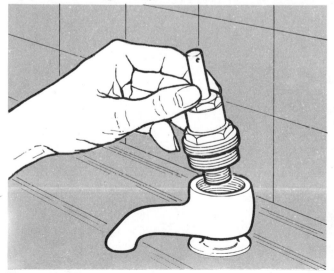

Fig. 52c Headgear of tap removed

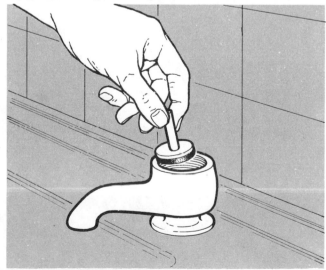

Fig. 52d Jumper and washer removed from tap

Fig. 53 Tap with capstan handle

can get at the headgear with the slim-jawed wrench, *fig. 52b*. Use the other wrench (padded) to grip the tap's body. Turn the headgear anti-clockwise to unscrew it. It lifts right out of the body of the tap, *fig. 52c*.

The headgear may also be difficult to shift. Sometimes a little drop of oil will help here too, but if it has become badly corroded with scale, considerable force may be needed to move it.

Be careful here not to exert too much force or you may wrench the tap away from its seating in the sink or basin. Use the other wrench as a lever to counter the muscle power.

With the headgear removed, the jumper of a cold water tap with the old washer attached, *fig. 52d*, will probably be seen sitting on the valve seating near the bottom of the body. (See also hot water taps section.)

If you intend only to renew the washer and not the jumper, unscrew the small nut which holds it to the jumper, release the washer, fit the new one and replace the nut. If it is impossible to shift the nut (it too may be scaled up) fit a new combined jumper and washer.

Replace this in the tap, screw back the headgear, tightly, and replace the easyclean cover (not too tightly).

Turn on the water (or release the arm of the ball float in the cistern) and your tap should now turn off without dripping.

CAPSTAN HANDLES

Before moving on to other types of tap, there are one or two points worth noting if this is your first attempt at rewashering. I suggested earlier that one of your wrenches should have slim jaws so that it can slip under the easy-clean cover.

This is not really essential if the handle of the tap is a capstan type, *fig. 53*. Capstan handles are secured by a small grub screw. Unscrew and remove this and the handle should come off the spindle by simply lifting it upwards.

Here again, however, you may have difficulty in shifting it. One answer is to protect the handle with a thick cloth and tap it gently from below with a hammer (a soft faced hammer is useful here). Be ready to grasp the handle or it may fly off and damage the sink or basin.

If this doesn't work, turn the tap on full, then insert an adjustable spanner on the spindle just above the easyclean cover. Turn the tap off. As you do so, the handle will be forced upwards and off. Don't use too much force or the cover may be damaged.

Having removed the handle and

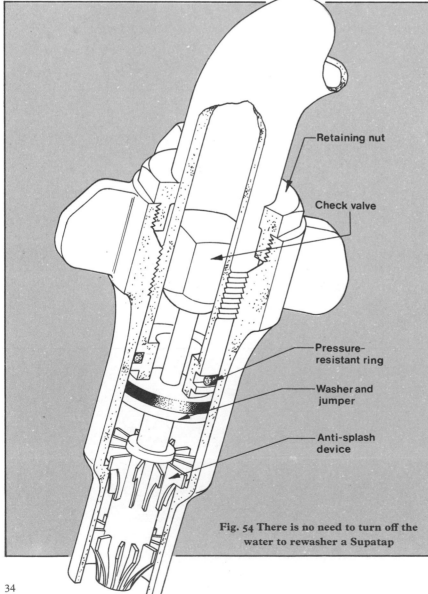

- Retaining nut

- Check valve

- Pressure-resistant ring

- Washer and jumper

- Anti-splash device

Fig. 54 There is no need to turn off the water to rewasher a Supatap

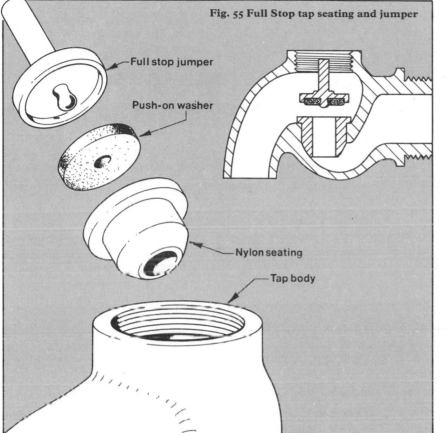

Fig. 55 Full Stop tap seating and jumper

- Full stop jumper
- Push-on washer
- Nylon seating
- Tap body

the cover, you are less restricted in your choice of wrench to shift the headgear of the tap.

Before replacing the cover and the headgear, grease the screw threads with Vaseline. Never overtighten a component.

REWASHERING SUPATAPS
Modern taps are different in appearance to the old patterns, but generally the rewashering procedure is similar. One exception to this, however, is the Supatap, *fig. 54*.

To rewasher this type of tap there is no need to turn off the water. First undo the retaining nut. Then turn the tap on and continue until the nozzle comes off. Don't be alarmed when the water starts to flow. This will soon stop when the check valve drops to seal the outlet.

If you then turn the nozzle upside down and tap it, the anti-splash device will fall out. Release the jumper and washer from the device; replace with a new jumper and washer combined.

Replace the anti-splash device in the nozzle and note that the nozzle is connected to the other part of the tap with a *left-hand* thread.

Water will flow again as the nozzle is replaced. When the nozzle is almost closed, tighten the retaining nut and close the tap fully.

Some taps have a combined handle and easyclean cover with a 'hot' and 'cold' indicator recessed in the cover. These indicators can be levered out gently with a penknife. Underneath will be a screw which holds the cover in position.

Undo the screw to remove the cover and the tap mechanism will be revealed. This should be similar to an ordinary tap.

HOT WATER TAPS
Rewashering a hot water tap is an operation similar to that with a cold water tap, but there is one difference. When the headgear is removed, the jumper and washer will probably still be in it instead of resting on the valve seating. Although you may be able to turn the jumper, it won't come away from the tap.

This is because it is pegged into the tap to ensure that, when the tap is turned on, the jumper and washer can be raised inside. Hot water pressure from the storage cylinder is normally too low to raise a jumper from the valve seating.

In these circumstances, the nut holding the washer to the jumper *must* be removed. Here you will probably have to apply a little oil to the nut and then give it time to soak through.

If the nut defies all efforts to shift it, the only thing to do is to prise the jumper out of the body of the tap with a screwdriver and replace it with a new jumper and washer combined.

PEGGING
Unfortunately, prising the jumper out will break the pegging, so the only way to ensure the new jumper will stay in the headgear of the tap is to burr the jumper so that it grips in the headgear.

This, however, should not be attempted until every effort has been made to unscrew the nut holding the old washer.

WORN SEATING
If a tap still drips after a new washer has been fitted, the valve seating of the tap may have become worn through constant use and age. There *are* reseating tools available, but the effort of doing the job is hardly worth while when you can buy a simple device to overcome the problem.

This is the Full Stop nylon washer and seating set combined, *fig. 55*, and costs only a few pence from ironmongers and some d-i-y shops.

All you do is drop the assembly into the base of the tap where the jumper

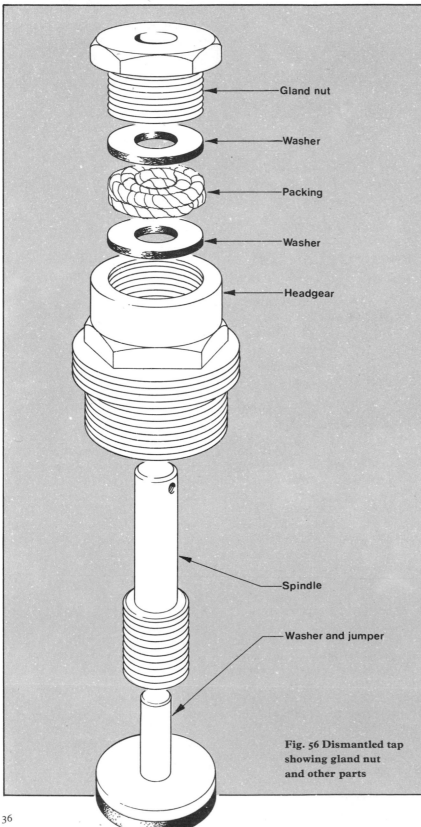

Gland nut

Washer

Packing

Washer

Headgear

Spindle

Washer and jumper

Fig. 56 Dismantled tap showing gland nut and other parts

of a cold tap normally rests. Make sure you get the correct size: there is one for kitchen and basin taps and another for bath taps.

LOOSE TAPS

Sometimes water will leak out of the top of a tap around the easyclean cover. This is a sign that the gland of the valve needs repacking or tightening.

The tap may turn too loosely in the hand when this fault develops. You may even be able to spin the tap with very little effort. Sometimes there will be the noise of water hammer as the tap is turned off.

To cure this, all that is necessary in some cases is to tighten the gland nut, *fig. 56*, about half a turn in a clockwise direction. Don't overdo it.

To get at the nut, you will have to remove the easyclean cover and take off the handle as described earlier. If there is no easyclean cover, the job is simpler, of course. There is no need to turn off the water to make this slight adjustment.

GLAND PACKING

There is, naturally, a limit to the number of adjustments that can be made to the gland nut. After a time, the gland will require repacking. If this is necessary, the water will have to be turned off.

Take off the handle and easyclean cover, then unscrew the gland nut and remove it and the fibre washer that lies beneath, *fig. 56*. Note that this is a different washer to those I have been talking about earlier.

Underneath the washer you should find the packing material which is usually hemp, but may be string or wool. Dig all this out (there may not be much left!) and beneath it you should find another fibre washer. Remove this too and renew it.

Repack the gland with hemp or wool smeared with Vaseline, pressing it down firmly, *fig. 57*. Renew and position the top washer and replace the gland nut – fingertight plus about half a turn.

Before replacing the handle and the cover, test the tap and adjust the gland nut as required. (You can turn the tap on by gripping the spindle with pliers.)

REWASHERING STOP-VALVES

As stop-valves inside the house are operated less frequently than other taps in the house, they don't need re-washering very often. Most internal stop-valves will resemble the one shown in *fig. 2*.

If one of these is fitted on the rising main pipe and needs rewashering, the main stopcock outside the house will have to be turned off first. If a stop valve on a supply pipe other than the rising main needs attention, turn off the water at the internal stop-valve on the rising main.

If you do not have a key to turn off the outside stopcock, you will have to get the local water board to do it, unless you make one as shown in *fig. 58*. But that, too, will depend on the type of stopcock fitted.

If your outside stopcock is as shown in *fig. 59*, then you will need a proper turnkey.

To rewasher the rising main stop-valve, first drain the water from the rising main pipe by turning on the kitchen cold water tap and the drain-cock incorporated with the stop-valve. (See *figs. 3* and *4*.)

If there isn't a draincock, have a bowl ready to catch any surplus water which runs out of the stop-valve.

The procedure now is similar to rewashering an ordinary tap except that there will be no easyclean cover to worry about.

If you rewasher a stop-valve on a pipe other than the rising main, this part of the plumbing circuit will need to be drained by running the appropriate taps until no more water flows.

Sometimes stop valves leak past their gland packing like other taps. The treatment here is similar to that with ordinary taps.

Check these valves regularly to ensure that they turn easily and don't become stuck. If they are stiff to turn, you can oil them occasionally.

Some internal stop-valves are called full-way gate valves, and they are usually to be found on pipes supplied by a storage cistern.

These valves normally need no maintenance apart from a slight adjust-ment of the gland packing nut – a rare requirement.

CHOOSING TAPS

Fortunately, taps have a long life and renewals are not often necessary. The appearance of a tap, however, has a great aesthetic value and modern taps are designed to please both in per-formance and in appearance.

There is an enormous variety of taps available these days and selection is not easy. So what should you look for when choosing new ones?

First of all, try the tap and make sure it turns smoothly and doesn't stick. Does it feel comfortable in the hand? If you are satisfied on these points the tap should be suitable

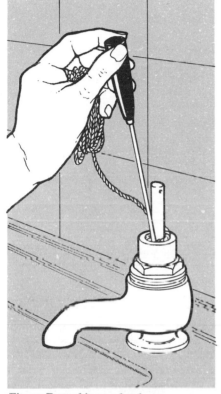

Fig. 57 Repacking a gland nut

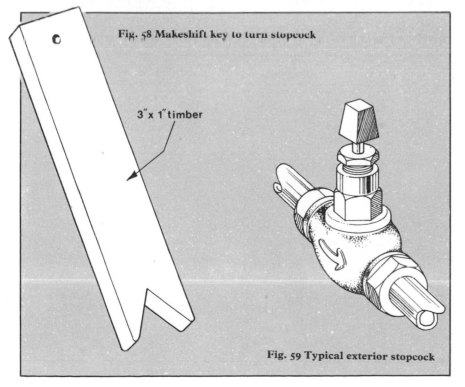

Fig. 58 Makeshift key to turn stopcock

3"x 1" timber

Fig. 59 Typical exterior stopcock

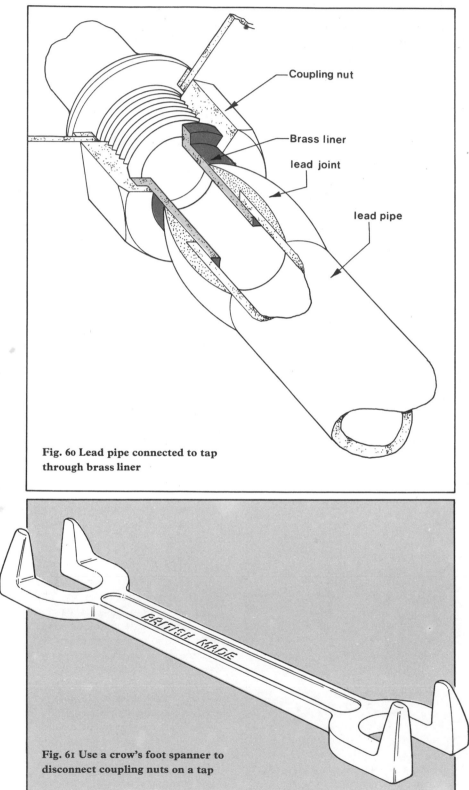

Coupling nut

Brass liner

lead joint

lead pipe

Fig. 60 Lead pipe connected to tap through brass liner

Fig. 61 Use a crow's foot spanner to disconnect coupling nuts on a tap

provided, of course, it is the right *type* for the situation.

If you are buying mixer taps, however, there is one very important thing to remember. A mixer should be designed to operate satisfactorily if, as is probable, the hot and cold water supplies are at different pressures. If the taps are not so designed, it is difficult to control the temperature of the water.

FITTING NEW TAPS

The job of replacing worn out or ugly taps is, frankly, one I would hesitate to recommend the unhandy handyman to attempt. Getting at the fitting can itself be a problem, and when removing old taps it is quite easy to damage a basin or sink.

So think twice before you attempt the job. Even if you think you can remove the old taps, make sure you know exactly how to fit the new ones. You may not be able to get hold of a plumber if you need one urgently to finish the job for you.

There are, however, many householders who have done this job themselves successfully, so if you feel confident, have a go.

The first thing to do, of course, is to cut off the water supply in one of the

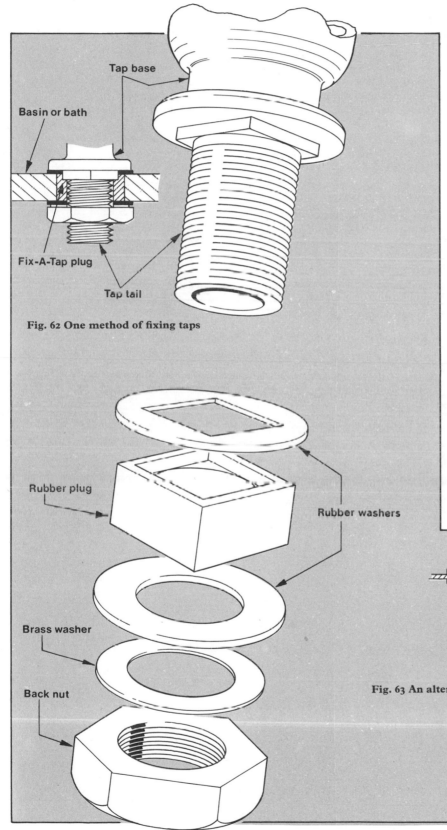

Tap base

Basin or bath

Fix-A-Tap plug

Tap tail

Fig. 62 One method of fixing taps

Rubber plug

Rubber washers

Brass washer

Back nut

ways I have described and to drain the supply pipes concerned. If the pipes are copper, the taps will be connected to them by a compression joint which can be unscrewed.

If the pipes are lead, they will be connected through a brass liner soldered to the pipe at the lower end and joined to the tap with a coupling nut, *fig. 60*.

CROW'S FOOT SPANNER
To turn these nuts, you will probably need a crow's foot spanner with right angle lugs, *fig. 61*. The design of the bath or basin may make it impossible to grip the nuts with an ordinary spanner.

The taps will probably be secured by nuts tightened against a bed of putty. If the plumber has done a good job when fitting the taps, the putty will be painted to give better adhesion.

This, then, will not be a simple joint to break, and if the taps are fitted to a small washbasin it may be easier to do the job (after the pipes are disconnected) if the basin is taken off the wall.

Taps have a square shank so that they do not twist in the holes, also square, provided in the fitting.

Therefore, if pressure is applied to

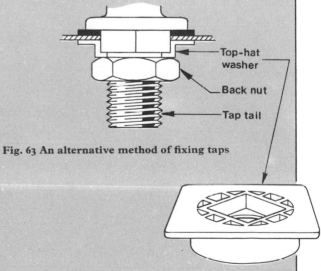

Top-hat washer

Back nut

Tap tail

Fig. 63 An alternative method of fixing taps

the coupling nuts without equal counter pressure being applied to the tap's body, the fitting may crack as the tap tries to turn in its hole.

To avoid this happening, remove the easyclean cover from the tap and apply the necessary counter pressure to the headgear with one hand, while unscrewing the coupling nut with the other.

PUTTY BED
To fix a new tap, first dab some enamel paint around the hole in the fitting. Then arrange a bed of metal casement putty around the flange of the tap and insert it in its hole.

Use a jointing compound to smear around the tap's threads and tighten up the coupling nut. Put some of the compound on the other end of the connection and refit the supply pipe.

Now this may sound fairly straight-forward, but you may find it difficult to make a joint with putty good enough to fix the tap firmly in its hole. The hole will be larger than the shank of the tap, thus giving a loose fit. Only practice can enable you to make a good job of the jointing.

THE EASY WAY
There is, however, an easier way of doing the job. You can buy what is known as a Fix-a-Tap set. All you do is to fit a Fix-a-Tap plug over the tap's shank and insert it in the hole in the fitting, *fig. 62*. The plug is then cut so that it protrudes slightly from the underside of the basin.

When the coupling nut is tightened, the protruding part of the plug is forced back into the hole. The result – a firm joint which can be taken apart at any time and reassembled.

You should be able to buy Fix-a-Tap sets, and the Full Stop fitting mentioned earlier, but if you have difficulty write to:

Nicholls & Clarke Ltd., 3 Shoreditch High Street, London, E1 (for Fix-a-Tap) and

Robert McArd & Co. Ltd., Crown Works, Denton, Lancs. (for Full Stop).

For thin basins and sinks you can use a special top-hat washer to ensure that the tap fits tightly in its hole, *fig. 63*. For ceramic basins, however, the Fix-a-Tap method is preferred.

Pipes and Joints

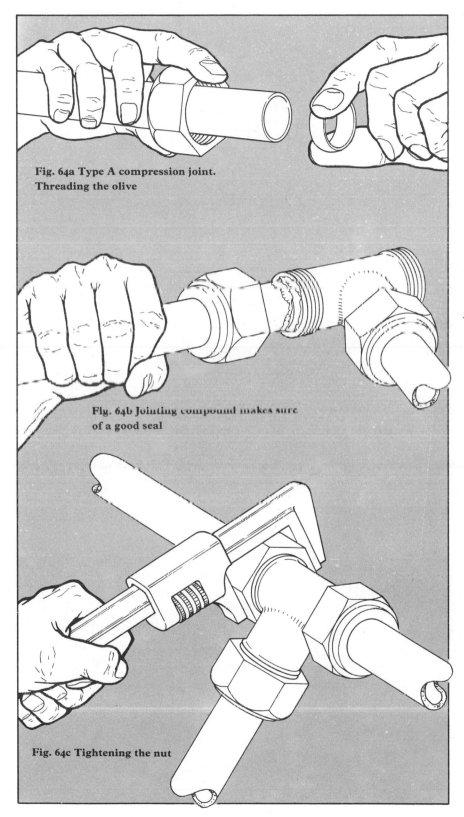

Fig. 64a Type A compression joint. Threading the olive

Fig. 64b Jointing compound makes sure of a good seal

Fig. 64c Tightening the nut

In the chapter on taps I mentioned a compression joint used to connect a tap to its supply pipe. Although for simple plumbing maintenance it is not vital to understand all the joints, pipes and fittings used in a domestic system, a little knowhow can be useful if, for example, you need to extend or replace existing pipework.

For ordinary domestic use, light gauge copper tubing is suitable and can be used as both hot water and cold water pipes. It can be cut simply, is available in small diameters and can be bent to shape without the need of costly tools.

As to joints, the simplest, I think, are non-manipulative (Type A) compression joints and fittings. A typical joint of this type is illustrated in *fig. 64a*.

COUPLING NUTS

These fittings can easily be disconnected at any time and reassembled. Threaded components are connected by coupling nuts. These screw down on to soft copper collars called olives or rings.

As the nuts are tightened, they crush the olives and make a watertight seal, but to make certain of this the connection can be smeared with a jointing compound.

A compression joint should be tight but should never be over-tightened. The drill is to twirl the coupling nut by hand as far as possible and then give it up to one full revolution of the spanner to tighten it up. If a joint weeps slightly, tighten it by degrees until it stops.

What sort of jobs would the average householder undertake which could involve the use of joints? One example is the fitting of an extension pipe to an existing one, the two pipes being the same diameter. If it is to be a straight run of pipe, you will need a simple compression coupling, *fig. 64a*.

The only tools needed for this job, apart from a spanner, are a hacksaw or tube cutter and two wrenches.

To extend the pipe, cut off the

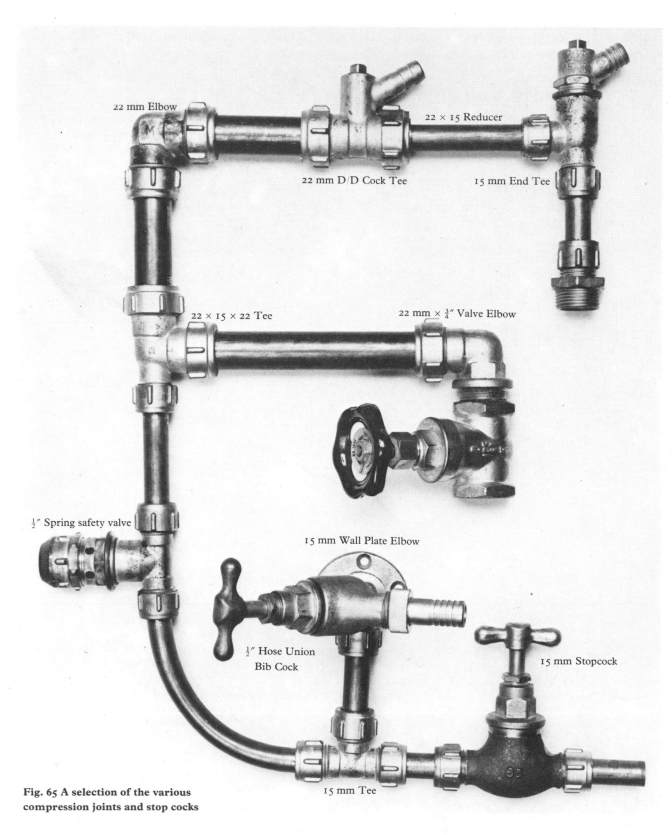

**Fig. 65 A selection of the various
compression joints and stop cocks**

22 mm Elbow

22 × 15 Reducer

22 mm D/D Cock Tee

15 mm End Tee

22 × 15 × 22 Tee

22 mm × ¾″ Valve Elbow

½″ Spring safety valve

15 mm Wall Plate Elbow

½″ Hose Union
Bib Cock

15 mm Stopcock

15 mm Tee

supply of water and drain the pipe. Cut
the pipe squarely if using a hacksaw –
after measuring carefully, of course –
and remove the internal burrs from
the pipe with a file or penknife, *figs. 69*
and *71*. If you use a tube cutter, *fig. 70*,
this may be fitted with a reamer which
will remove the burrs.

Loosen the nut on the compression
joint and insert one end of the cut pipe
into it as far as it will go, *fig. 64b*.
Tighten the nut with the wrench,
fig. 64c.

Repeat the performance with the
other end of the cut pipe and your
compression joint is made.

FITTING A STOP-VALVE

A compression joint can also be used
to fit an internal stop-valve on a supply
pipe.

First turn off the water supply and
drain the pipe. Measure the distance
between the shoulders, or stops, on
the inside of the threaded ends of the
compression joint and cut out the
correct amount of piping.

Take great care when cutting the
pipe. Cut out too little rather than too
much; you can trim a little off, but
you can't put any back on!

If you *should* make a mistake and cut
the pipe too short, you can, of course,
fit a compression coupling joint to fill
the gap, but this would be an un-
necessary expense to incur simply
through carelessness.

Burr off the ends of the cut pipe,
slip on the coupling nuts and then the
olives to each end of the pipe and insert
the pipe into the valve as far as it will
go. You will feel it hit the stops when
it is in to its full extent.

Note that the arrow on the stop-
valve should always point in the
direction of the water flow.

Screw the nuts on to the stop-valve
and tighten them up over the olives.

TYPE B FITTINGS

The compression joints mentioned
above are *non*-manipulative joints.
Manipulative fittings are slightly
different and are designed mainly for

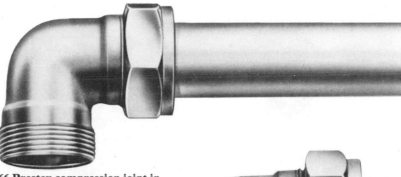

Fig. 66 Prestex compression joint in
assembly order. The tube should enter
the fitting as far as the stop

Fig. 67 The two-point seal when the cone
is compressed

Fig. 68 A steel drift to open up the end of a pipe

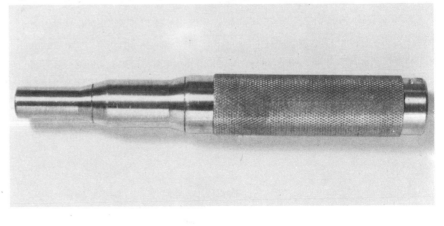

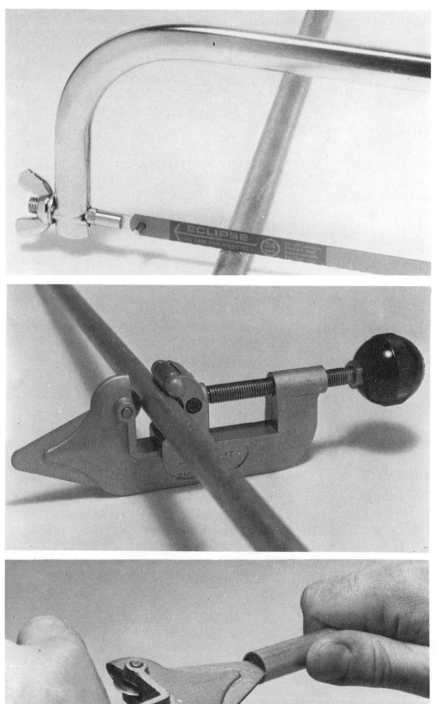

Fig. 69 Cutting a pipe with a hacksaw

Fig. 70 Tube cutter

Fig. 71 Removing burrs

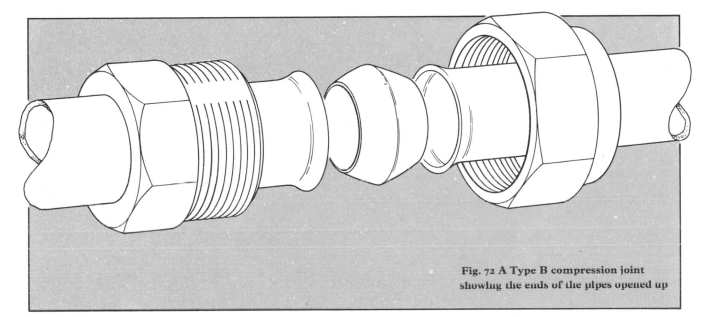

Fig. 72 A Type B compression joint
showing the ends of the pipes opened up

work underground where the joint will
be inaccessible. Once this type of joint
has been formed, it cannot be pulled
apart.

Manipulative compression joints are
known as Type B fittings. When using
these joints, the shape of the end of the
pipe has to be altered so that it can be
gripped by the fitting.

One way to do this is to open up the
end of the pipe by hammering a steel
drift, *fig. 68*, into it.

To make a Type B joint, the holding
nut is removed from the body of the
joint and slipped over the end of the
pipe before it is manipulated.

The next step depends on the
method used. A steel drift is used to
manipulate Prestex, Conex and Kontite
joints. For Kingley fittings you use a
swaging tool.

So, depending on which fitting has
been chosen, either open the end of the
pipe with the steel drift, or put the
swaging tool in the end and turn its
handle. Grip the pipe firmly while
turning.

In the case of the Kingley fitting, the
swaged (or ridged) end of the pipe is
now pushed into the fitting as far as
possible, and the nut tightened against
the swage.

In the other method, the flared end

Fig. 73 Cleaning up the end of a pipe with wire wool
before making a capillary joint

of the pipe is positioned on the
shoulder of the joint's body and the
nut tightened.

CAPILLARY JOINTS

Another type of joint more likely to be
of interest to a handyman than Type B
fittings is the soldered capillary joint.
This, of course, involves the use of a
blowlamp or blowtorch.

If this is the sort of joint you prefer,
it should be borne in mind that the
flame from a blowtorch can be quite

fierce! When working in confined
spaces it is only too easy to let the
attention wander and direct the flame
on to combustible material such as
timber framework, rafters or skirting.
So take care if you use this method.

Personally, I would suggest a blowtorch rather than a blowlamp for a
beginner.

An integral ring joint is the simplest
form of capillary joint for a d-i-y man
to make. This has a ring of solder
inside.

45

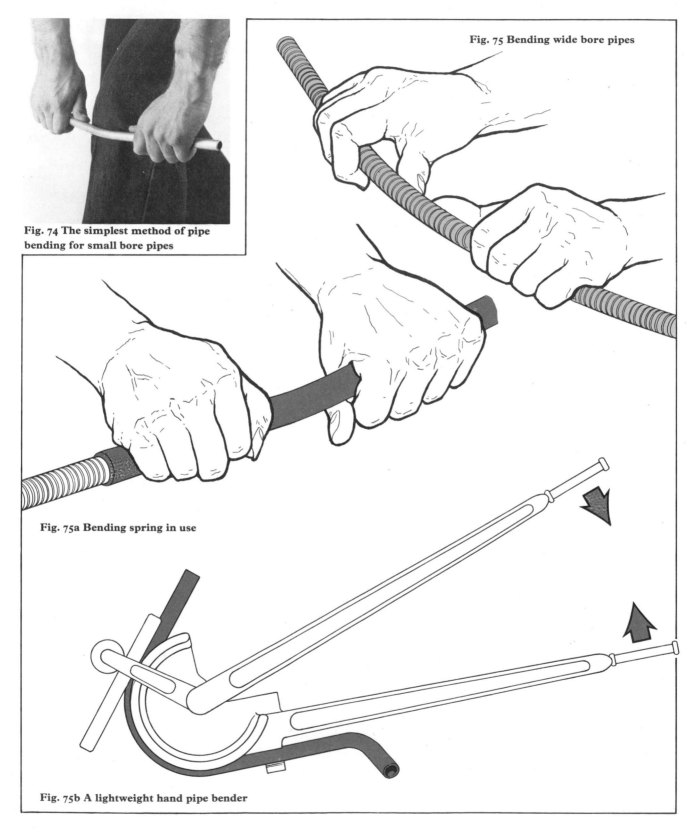

Fig. 74 The simplest method of pipe bending for small bore pipes

Fig. 75 Bending wide bore pipes

Fig. 75a Bending spring in use

Fig. 75b A lightweight hand pipe bender

An example is shown in *fig. 76*. These are sometimes called Yorkshire joints.

To make this type of joint you cut the pipe as before and remove the burrs from its ends. You then have to clean the outside of the pipe and the inside of the capillary fitting with fine grade emery paper.

RING OF SOLDER

Now apply some flux to both the cleaned areas and slip the pipe into the fitting *as far as possible*. Direct the flame of the blowtorch on to the fitting until the solder starts to run.

When you can see an even ring of solder all the way round the end of the fitting, the joint is complete and should now be left to cool off.

There are other types of capillary joint, including the Endfeed, *fig. 77*, and Konsilet fittings. In these, the solder is fed into the mouth of the fitting using solder wire.

After cleaning the fitting and the pipe, and inserting it into the fitting as before, you direct the flame of the blowtorch onto the body of the fitting and gradually move it towards the mouth.

At this point, you then apply solder wire to the mouth until the ring of solder is complete as before.

STAINLESS STEEL TUBE

Although copper pipe is widely used in domestic plumbing systems, there are alternatives.

One is stainless steel tube which has been more widely used in the past few years. This material is harder than copper and is not so easy for an amateur to bend by hand.

However, its rigidity means that it requires less support than copper pipe and it can withstand greater water pressure.

It is possible to use Type A compression joints with stainless steel tube and it can be cut in the same way as copper. You cannot, however, use Type B compression joints with this material. Soldered capillary joints can

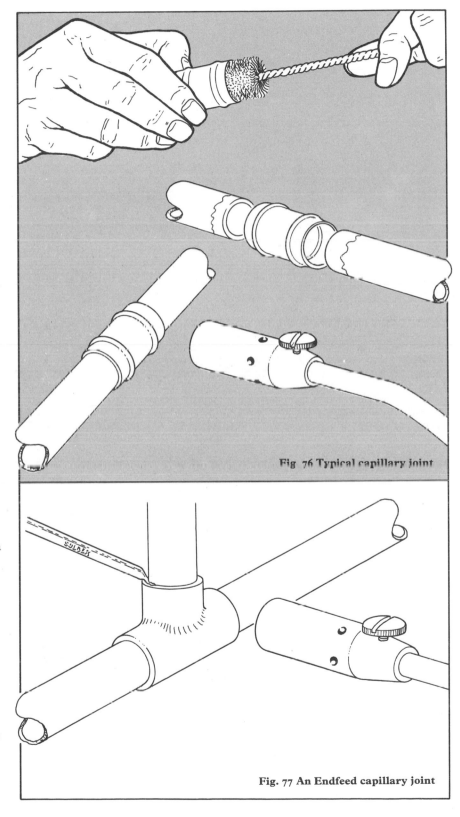

Fig 76 Typical capillary joint

Fig. 77 An Endfeed capillary joint

Fig. 78 A selection of typical capillary joints

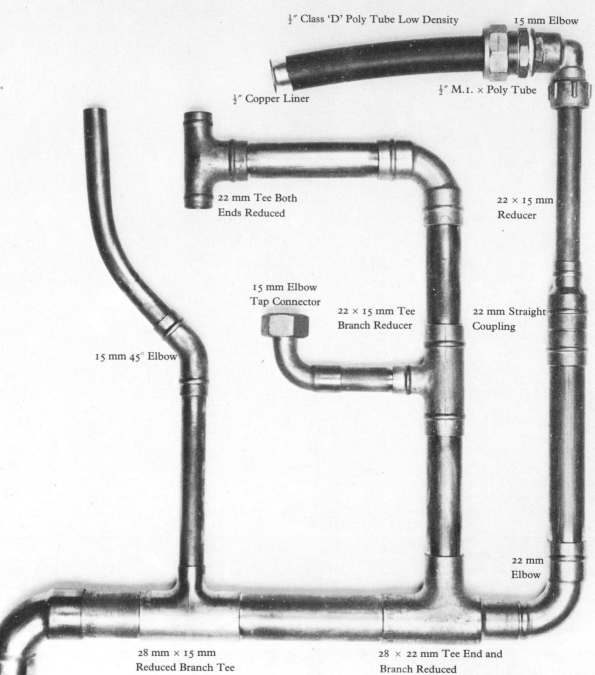

½″ Class 'D' Poly Tube Low Density

15 mm Elbow

½″ Copper Liner

½″ M.I. × Poly Tube

22 × 15 mm Reducer

22 mm Tee Both Ends Reduced

15 mm Elbow Tap Connector

22 × 15 mm Tee Branch Reducer

22 mm Straight Coupling

15 mm 45° Elbow

22 mm Elbow

28 mm × 15 mm Reduced Branch Tee

28 × 22 mm Tee End and Branch Reduced

Conversion Table. Inches to millimetres

in	mm	in	mm	in	mm
$\frac{1}{32}$	0.794	$\frac{9}{32}$	7.144	$\frac{5}{8}$	15.875
$\frac{1}{16}$	1.588	$\frac{5}{16}$	7.938	$\frac{11}{16}$	17.463
$\frac{3}{32}$	2.381	$\frac{11}{32}$	8.731	$\frac{3}{4}$	19.050
$\frac{1}{8}$	3.175	$\frac{3}{8}$	9.525	$\frac{13}{16}$	20.638
$\frac{11}{64}$	4.336	$\frac{7}{16}$	11.113	$\frac{7}{8}$	22.225
$\frac{3}{16}$	4.763	$\frac{1}{2}$	12.700	$\frac{15}{16}$	23.813
$\frac{1}{4}$	6.350	$\frac{9}{16}$	14.288	1 in	25.400

Note: metric-sized pipes are measured across their *outside* diameters. Imperial-sized pipes are measured across their *internal* diameters. E.g. the equivalent of a $\frac{1}{2}$ in pipe in metric size is 15 mm.

be used, but a special flux which is based on phosphoric acid is required.

POLYTHENE TUBE
Polythene tube can be used for cold water pipes, but not for hot because it softens when heated and will melt at about 230° Fahrenheit. Possible uses for it in a domestic plumbing system include waste pipes, cold water pipes in the loft and to supply taps at the end of a garden.

This material comes in normal gauge and heavy gauge. Most water boards insist on heavy gauge if the pipe concerned is to be under mains pressure, so this is a point to check with your board before changing over any part of the system's pipework to polythene.

Non-manipulative compression joints can be used for joining, but when ordering the joints, state the tube's internal diameter and its gauge. For many joints you will need a size larger than the size of the tube.

RIGID PVC
Another type of pipe more widely used nowadays in *cold* water plumbing is rigid pvc, which is self-supporting. Its rigidity makes it difficult to bend, however, and it is easier to avoid bends by using elbow or similar joints instead, where necessary.

One point to watch when using this tube is to avoid running it close to hot water pipes. Keep the cold water and hot water pipes at least 3 in apart.

METRICATION
Like other industries, plumbing has been affected by the gradual change-over to metrication. The main concern of home plumbers in this respect is the altered sizes of pipes and joint fittings. These are now sold in metric sizes.

One question likely to be asked by home plumbers is: how can an imperial-sized pipe be extended if the compression joints now available are in metric sizes? No great problem is involved, however.

To take pipe sizes. Probably the only ones likely to be used by a handyman are $\frac{1}{2}$ in, $\frac{3}{4}$ in and 1 in. The metric equivalents of these sizes are 15 mm, 22 mm and 28 mm.

There is very little difference in size between 15 mm and $\frac{1}{2}$ in pipes in actual measurement. The reason that they are similar is because they are not measured in the same way.

NO NEED FOR ADAPTORS
For imperial measurements the quoted size refers to the *internal* diameter of the pipe, whereas metric measurements refer to the *external* diameter; hence the apparent discrepancy in measurement.

Therefore, both sizes can be used with 15 mm compression joints without the need for adaptors. The same thing applies to 1 in and 28 mm size pipes.

The $\frac{3}{4}$ in and its 22 mm metric equivalent are not quite the same size. Nevertheless, you can use 22 mm compression fittings with $\frac{3}{4}$ in pipe provided you use $\frac{3}{4}$ in cap nuts and olives (rings) instead of the metric-size cap nuts and rings.

CONVERSIONS
Capillary soldered fittings, however, can cause problems when metric-size pipes need to be fitted to imperial sizes. Capillary fittings are more exact than compression types and the two sizes (imperial and metric) are not interchangeable.

However, the problems are not insoluble. You can buy conversion items which have *inlets* in the old imperial sizes and *outlets* in the new metric sizes.

That is one solution. Another is to fit a compression joint at the point where the new size pipe is to connect to the old. For any subsequent joints along the run of new pipe, metric capillary joints can be used.

You will find that there is an enormous variety of compression and capillary fittings available – bends, straight couplings, elbows, tees and so on.

If you plan to carry out a lot of pipework, I suggest you send for catalogues to the following manufacturers. You will find them extremely useful.

Compression fittings: Conex Sanbra Ltd., Great Bridge, Tipton, Staffs. DY4 7JU. Peglers Ltd., St Catherine's Avenue, Doncaster, Yorks. DN4 8DF. **Capillary fittings:** Yorkshire Imperial Metals Ltd., P.O. Box 116, Leeds, LS 1RD. Kay & Co. (Engineers) Ltd., Blackhorse Street, Bolton, Lancs. King's Langley Engineering Co. Ltd., King's Langley, Herts.

Frost Precautions

As a nation we are now becoming more conscious of the vital need to insulate our homes to conserve heat and thus save money on fuel.

That is a good thing, for a well insulated house will also go far to protect the plumbing system against frost damage – provided that the heat which is circulated can reach vulnerable pipes and plumbing equipment.

Let us suppose you have taken precautions to insulate your home. In the loft you have probably laid vermiculite or other granules or spread glass fibre blanket material between the the joists. By doing this, you have ensured that valuable heat will not rise through the first floor ceiling into the loft and be finally dissipated through air spaces in the roof itself.

But what of the plumbing which is up in the loft? Are all the pipes and the cold water storage cistern thoroughly lagged to prevent icy draughts from freezing them up in severe weather? That, oddly enough, is something a great many people overlook.

WATCH THOSE BENDS

All plumbing materials in the roof space should be adequately lagged. Bends in pipes should receive particular attention, especially those which are situated in remote corners of the loft far away from any warmth which may reach the area from below.

The cold water storage cistern should be lagged unless it is a polythene type. It should not be necessary to lag these cisterns but they, like any other type (galvanised steel or asbestos cement), should have a dustproof cover.

This is necessary not only as a precaution against freeze-ups, but, as mentioned earlier, to prevent the water from being contaminated by insects and debris which may drop into it from the roof.

EXPANSION PIPE

Make sure that the expansion pipe over the cistern is also lagged. A hole can be made in the cover on the cistern so that any water which runs from this

Fig. 79 Blanket-type lagging for the loft

Fig. 80 Lagging jacket for the hot water tank

pipe can enter it. If you fit a plastic funnel in the hole the water will not splash on to the cover.

When insulating the floor of a loft, by the way, do not lay insulation material *under* the storage cistern and don't lag the base of it. By leaving these parts uninsulated, a little warm air from below will then be able to rise and help to prevent the base of the cistern freezing up.

When lagging the cistern, don't overlook any stop-valves which may be fitted to the outlet pipes. Lag these thoroughly, leaving just the handles visible.

BOARDING THE LOFT

Insulating the plumbing system in a loft is not a comfortable job, but it is one which no householder should neglect. You can make the job easier by screwing boards across the rafters, where necessary, so that you can walk about without fear of slipping a foot through the ceiling of the room below!

A good place to start boarding is on the route from the trap door to the cold water storage cistern. You never know when you may have to dash into the loft and attend to a faulty ball-valve.

Not all the area in the roof space normally needs to be boarded over; catwalks as in *fig. 81* will probably suffice. Much will depend, of course, on the layout of the pipes.

Incidentally, I suggest screwing the boards down rather than nailing them. If you use screws, any cables and pipes hidden under the boards will be more easily accessible in future. If any pipes are buried under the boards, make a note of what they are for quick future reference.

OVERFLOW PIPE

While lagging the pipes in the loft, don't overlook the overflow pipe from the cold water storage cistern. If this should freeze up, and the ball-valve should happen to stick in the open position, the result could be disastrous. This open-ended pipe is an invitation to icy blasts to whistle up it

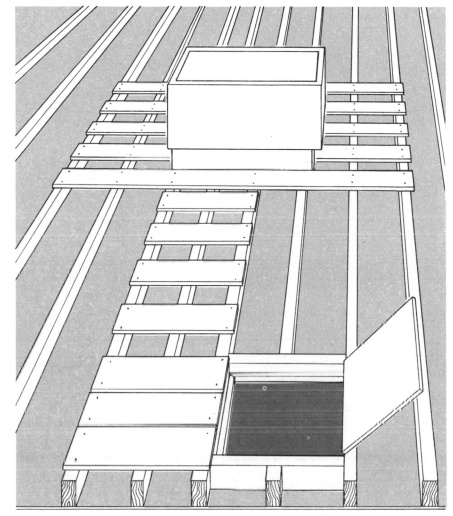

Fig. 81 Boards over the rafters in the loft provide easy access to the plumbing

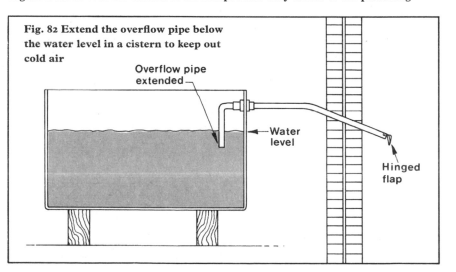

Fig. 82 Extend the overflow pipe below the water level in a cistern to keep out cold air

Overflow pipe extended

Water level

Hinged flap

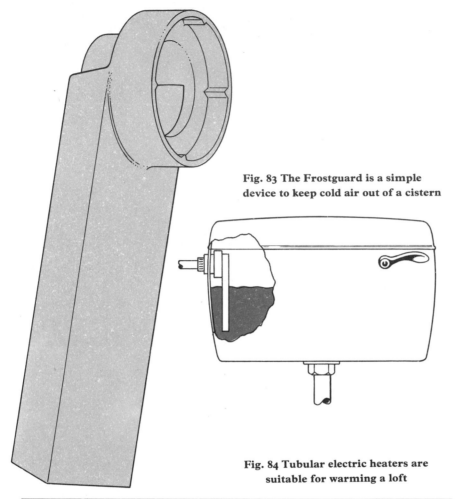

**Fig. 83 The Frostguard is a simple
device to keep cold air out of a cistern**

**Fig. 84 Tubular electric heaters are
suitable for warming a loft**

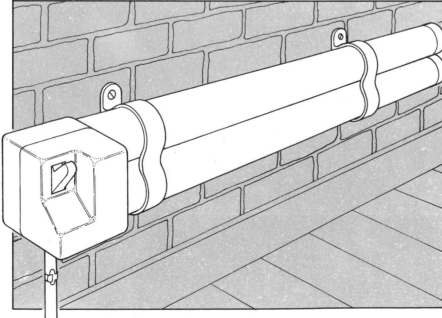

and freeze the water in the cistern, but
open ended it has to be.

There are a couple of ways of
avoiding a freeze-up. One is to arrange
the pipe so that its internal end is
continued into the cistern and terminates below the level of the water,
fig. 82.

Any draughts entering the overflow
pipe will then meet a barrier of water.

Another method is to fit a small
hinged flap (if you can get one) to the
external end of the overflow pipe,
fig. 82. As cold blasts of air hit the flap
from outside, they keep it in the closed
position.

If you choose this method, keep a
watchful eye on the flap and make sure
it is kept free of dirt and corrosion.
Otherwise it may jam open and become
temporarily useless until released.

DANGER
Another reason I don't favour this
method is because there is always the
danger of the cistern overflowing and
causing icicles to form on the overflow
pipe. If this ice jammed the flap in the
closed position, the pipe would not be
able to do its job if the cistern over-
flowed again. Flooding in the house
would be the result.

An alternative method to extending
the overflow pipe into the cold water
cistern is to fit a Frostguard. This
simple device, shown in *fig. 83*, simply
screws on to the *internal* end of the
pipe.

USING HEATERS
When you are satisfied that everything
has been lagged in the loft, you will
have done much to ensure a winter free
of worry from frost damage in one
area at least. But it is by no means
certain that your lagging precautions
will prove effective against a *very* severe
frost.

All that lagging can do is to keep
frost at bay for a time; it cannot, in
itself, provide extra warmth for the
plumbing installation.

Therefore, a wise householder will
consider introducing some form of

heating in the loft, especially if the house is an old one and the roof is not lined.

I think the only suitable and safe form of heating for this purpose is electricity, and I suggest one or more tubular heaters, *fig. 84*. These operate at black heat, rather like a radiator.

A radiant fire is not to be recommended for this purpose. Neither is a paraffin heater, no matter how safe it is thought to be.

Tubular heaters are sold in various lengths and operate at a rate of about 60 watts to 1 ft. A 6 ft length (360 watts) is suitable for many requirements, but more heat than this will be needed in most lofts.

These heaters, by the way, should never be run off a lighting circuit as this could cause an overload. Operate them from earthed socket outlets which have been correctly wired.

Another suitable way to provide warmth in the loft is to use a fan heater. This will warm up a loft rapidly.

WARNING

Having said that, I must, however, add a warning when using any sort of loft heating. This is a place where all sorts of things get thrown up from below. It is the easiest thing in the world to toss up an unwanted rug or carpet so that it becomes accidentally draped over a fire up there.

Later, when the fire is turned on, possibly from a switch on the landing, the carpet may well have been forgotten. The heater can then overheat and cause a fire in the loft. This will probably not be noticed until it is too late to prevent extensive damage being done.

One more thought on loft insulation before we move on to the rest of the plumbing system. If you want to keep the loft at a reasonable temperature, consider insulating the rafters instead of the floor. If you can see daylight through the tiles of an unlined roof, that is one spot where your precious warmth will make its escape.

One suitable form of roof insulation

is waterproof paper or sheet aluminium foil pinned to the rafters with generous overlaps at the joints.

When all the plumbing in the loft has been insulated, the danger of frost damage to the system will have been reduced but not dispelled altogether. There are other parts which are vulnerable and this is especially true where older houses are concerned.

CHECK THE WHOLE SYSTEM

It is worth carrying out a check of the complete installation, starting from the stopcock outside the house.

As this will be (or should be) buried at least 30 inches below ground level, it is unlikely to be affected by frost unless the cover of the stopcock is loose or inadequate, or the ground level has been lowered since the stopcock was installed, which again is unlikely.

You won't be able to get at the stopcock to lag it, of course, so the only alternative is to drop some vermiculite granules (loose fill insulation) down the hole until only the top of the stopcock is exposed, so that it can be turned off when required.

THE RISING MAIN

From that point, the rising main pipe should already be protected by 30 inches of earth until it emerges inside the house.

Commonsense will tell you whether this pipe needs lagging at this point. If the pipe rises in a warm kitchen against an *internal* wall (as it should), no lagging should be necessary to protect it against frost, but lagging is advisable if the pipe rises against an *external* wall.

CONDENSATION

There is, however, one other very sound argument for lagging the rising main pipe in a kitchen. Have you ever noticed how wet this pipe becomes as a result of condensation? You can prevent this by lagging the pipe thoroughly so that no part of it is exposed.

Fig. 85 Fix plastic foam lagging material with adhesive tape

This is especially important if the pipe is fixed close to an external wall. Get the lagging material *behind* the pipe as well as in front to insulate it.

All these remarks apply equally to other pipe runs in the house which may be affected by condensation. Incidentally, it will be found that painted pipes are sometimes less affected by condensation than unpainted ones.

The appearance of a pipe in the kitchen or elsewhere is a consideration, of course, and some ordinary lagging material is not exactly a joy to the eye! Less unattractive – and certainly very effective – are lengths of foam plastic lagging material which are simply slipped over the pipes, *fig. 85*, and secured with adhesive tape.

This form of insulation will not only help to protect a cold water pipe, but can also be used to lag hot water supply pipes in order to reduce heat losses.

INTERNAL TOILETS

Another spot vulnerable to frost is, of course, an internal toilet cistern and its supply pipe. In a house which is

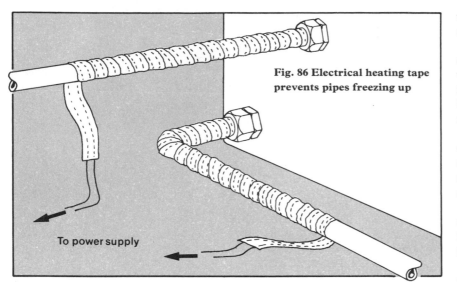

To power supply

Fig. 86 Electrical heating tape prevents pipes freezing up

constantly warm the danger of a freeze-up here may be remote, but in a cold house the danger is very real.

The overflow pipe of the toilet cistern can, as stated earlier, admit icy draughts, and although these cisterns are fitted with covers, in a severe winter the water in them could freeze.

The problem can again be overcome by turning the overflow pipe's internal end into the cistern below the water line or by fitting a Frostguard as suggested for the cold water storage cistern.

If you fit a Frostguard, by the way, make sure that the water level does not then rise above its normal line, and ensure also that you can fit it without it fouling any part of the mechanism in the cistern.

OUTSIDE TOILETS
Outside toilets pose special problems. Many are very draughty places, sometimes with no light and with great gaps between the foot of the door and the doorstep.

There is little point in carrying out anti-freeze precautions in such a toilet until it has been thoroughly draught-proofed. Even when that is done, the fact that the room is unlikely to be heated will make its plumbing system vulnerable to frost.

So try to introduce some heat – even

the heat from a 100 watt bulb placed near the cistern's ball-valve will be a help – but avoid radiant fires and paraffin heaters because of the fire risk.

There are more advanced methods of protecting exposed pipes from frost apart from lagging. One which is very popular is electrical heating tape. This is simply wound around the pipes, *fig. 86*. This tape can be plugged into a socket outlet and turned on whenever a frost is forecast.

INSULATION MATERIAL
Most insulation material is available from builders' merchants.

For wrapping round cold water storage cisterns you can get glass fibre blanket or expanded polystyrene tank lagging sets.

Insulation jackets are available for hot water cylinders, and pipe lagging material is sold in various forms.

OTHER PRECAUTIONS
There are other precautions which should be taken whenever frosty weather is likely. Briefly, they are:

1. If you have any dripping taps, get them attended to before winter sets in.

2. Don't leave a tap to run all night during frosty weather in the hope that this will prevent it freezing up. There is a danger during severe frost of the

water freezing in the waste pipe outside.

If this happens and the tap is allowed to run on, the whole of the waste pipe could eventually become frozen up. The result – a flood of water in the room where the tap is left running.

3. When leaving the house unoccupied for a few days during a cold spell, drain the whole of the plumbing system. Then flush the toilet cistern and put a handful of salt in the pan.

Remember, when you return, to refill the system before lighting the boiler! And make sure the system *is* filled before doing so.

These precautions are perhaps not strictly necessary in a house with a well-appointed central heating system which is automatically controlled. But the plumbing system should nevertheless be adequately lagged.

BURST PIPE
If, after taking all possible frost precautions, a cold water pipe bursts, the first thing to do is to stop the supply of water to the circuit involved. Do this *before* you look for the burst.

Should the burst occur in the rising main, turn off the internal stop-valve or the stopcock in the pavement.

If there is a burst in a supply pipe leading from the cold water cistern, turn off the stop-valve on that pipe.

If the supply pipe to the hot water cylinder bursts, the stop-valve in this case may also be near the cistern or possibly in the airing cupboard near the cylinder.

PLUG THE OUTLET
If no stop-valves are fitted on the pipes leading from the cistern, the outlet to the supply pipe affected can be plugged from inside the cistern with a thick cloth wrapped around a length of broomstick. Then tie up the ball-valve of the cistern so that no more water can pass into it.

The kitchen cold water tap can still be used in this case provided, of course, that it is supplied direct from the main.

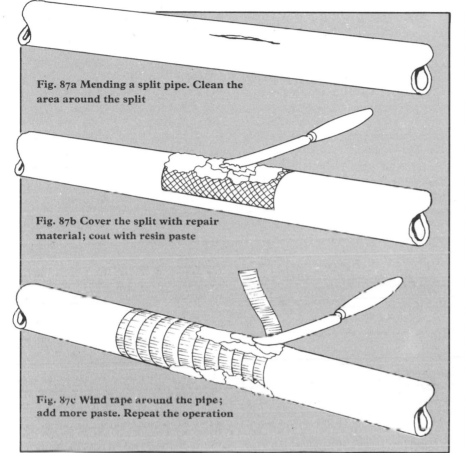

Fig. 87a Mending a split pipe. Clean the area around the split

Fig. 87b Cover the split with repair material; coat with resin paste

Fig. 87c Wind tape around the pipe; add more paste. Repeat the operation

If the stopcock in the pavement, or the internal stop-valve on the cold water supply pipe to the hot water cylinder, is turned off, you should also, as a precaution, turn off the boiler or, in the case of a solid fuel type of boiler, remove the fuel with a shovel.

In an emergency such as this, knowing your plumbing system is half the battle. As soon as water starts to drip from a leaking or burst pipe, you may be able to stop this turning into a flood if you can quickly locate the fault and isolate that part of the system. But first, concentrate on cutting off the water supply.

I repeat what I said earlier: get to understand your system thoroughly. A good time to do this is in the summer when the boiler is not operating. Then there will be no fear of it exploding if you accidentally drain it of its supply of water.

MENDING LEAKS

Having found the leak, what you do next depends on the type of pipe involved. The water may be coming from a compression joint in a run of copper pipe which has simply been forced apart by the frost. If this is the case, then you are lucky! A turn of the nut with a spanner may do the trick.

If, however, you have lead pipes and one of these has split, the correct method of repair is to cut out the affected length and replace it with a new length.

This, however, involves making wiped soldered joints which I consider are not strictly do-it-yourself jobs, although I hasten to add that many a d-i-y man *has* tackled the job successfully.

TEMPORARY REPAIRS

With the advent of synthetic resins, however, it is possible to make temporary repairs to split pipes which will at least enable them to be used until the job can be done properly. In some cases, the repair may well last for years.

Isopon and Plastic Padding are two types of repair kit you can get easily. It is a good plan to keep one handy.

Figures 87a to *87c* illustrate how a repair can be made.

Clean the pipe around the split and remove any paint. Bridge the split with glass fibre mat. Apply the resin (mixed with hardener) over the area; wind glass fibre tape round the pipe. Brush on another layer of resin/hardener mix. Repeat the process if necessary.

The method may vary according to the product used.

If a pipe has frozen up but has not burst, locating the frozen section will obviously be a little more difficult, but it needs to be found quickly before the freeze-up spreads to other sections of the pipe.

By turning on various taps you should be able to locate it fairly soon. One of the most likely places is in the loft and it could be that the rising main is the one affected.

When you have located the trouble, try applying some cloths, dipped in hot water and then squeezed out, to the suspect pipe. You may not need to treat more than one or two spots on the pipe as it will itself conduct the applied warmth along its length.

WATCH THE BENDS

Other pipes which are most likely to freeze up will be the supply pipes leading from the cold water storage cistern, especially if they should happen to run near the eaves.

Go for any bends in the pipes. This is where ice can form and quite often these bends are situated in awkward spots difficult to reach comfortably.

These bends are also a problem to lag thoroughly, so if you spot an exposed section of pipe where the lagging has, perhaps, slipped, this may be where the freeze-up has occurred.

For reaching awkward corners try

using a hair drier or a fan heater with the warm air directed at the suspect spot.

Quite often a warm (not hot) water bottle laid on the pipe will do the trick, too.

For pipe runs which are exposed to draughts, normal lagging may not be enough to prevent them freezing up. It is these, especially, which will benefit from the electrical heating tapes I have already mentioned.

If these are not readily available, Hotfoil Ltd., Heathmill Road, Womborne, Wolverhampton, can tell you all about them.

KEEP THE BOILER BURNING

When the weather is frosty, never put the boiler out at night because you fear that it might explode. Even if you have no central heating and use the boiler simply to heat water for normal domestic use, the heat the boiler distributes helps to keep the temperature of the house above freezing level.

All the pipes between the boiler and the hot water cylinder will be kept warm and will themselves act as miniature radiators. If your cold water storage cistern is positioned immediately above the hot water cylinder – an

excellent arrangement – this will receive its quota of warmth from the vent pipe.

The danger of a boiler explosion is more real if the boiler is used after being left unlit during a long period in frosty weather and the plumbing system has been frozen up.

Ice can form in the cold water supply pipe to the hot water cylinder, in the hot water system's vent pipe, or even in the flow and return pipes which run between the cylinder and the boiler, *fig. 88.*

Ice in any of these pipes will seal the hot water system which should normally be open and under pressure from the atmosphere. Therefore, when the boiler is relit, the water in it will not be able to circulate and expand. So pressure in the boiler increases until finally it is released by the whole system exploding!

CYLINDER COLLAPSE

This, luckily, does not happen very often, but there is another danger which arises more frequently in cold weather – cylinder collapse. This can happen on a freezing night if the boiler is allowed to go out and the hot water system cools down.

Ice can then collect in the vent pipe and in the pipe which supplies cold water to the cylinder. This causes the system to be sealed off as before, but now the system is cooling down.

As it does this, the water contracts. This causes a vacuum to form in the system and the cylinder (which is not designed to stand up to external pressure) will collapse. This collapse may take place when a hot water tap is turned on.

So be warned. Don't let the boiler go out in freezing weather.

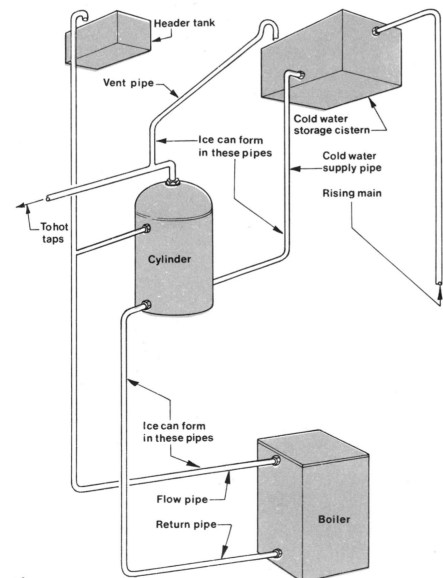

Header tank
Vent pipe
Ice can form in these pipes
Cold water storage cistern
Cold water supply pipe
Rising main
To hot taps
Cylinder
Ice can form in these pipes
Flow pipe
Return pipe
Boiler

Fig. 88 Pipes vulnerable to frost

Clearing a Blockage

One of the most irritating things that can happen to a housewife is a blockage in the waste pipe of the kitchen sink. This is often caused by something becoming stuck in the trap underneath. This may be a build-up of tea leaves, pieces of vegetable or congealed fat – or a combination of all three.

In many instances these blockages can be cleared quickly by using a rubber suction cup, *fig. 91*. Run water into the sink, seal off the overflow outlet to exclude air and then pump the suction cup rapidly up and down over the plughole.

If this doesn't work, the blockage may be too solid to shift in this way. But repeat the treatment a few times before admitting defeat. Then examine the trap under the sink.

If this is a bottle trap, *fig. 100*, unscrew its bottom section. If it is a U

Fig. 91 A suction cup is useful for clearing sink blockages

trap, *fig. 99*, unscrew the inspection plug in the bend.

Place a bucket or bowl under trap. Poke a length of springy wire into the hole in both directions.

If you manage to remove the obstruction, flush through with boiling water before replacing the bottle trap section or inspection plug.

BLOCKED BEND

If you don't find the object, it may be lodged in the bend of the waste pipe where it passes through the wall to the outside gully. Try poking a cane up the pipe from outside or try to reach the blockage from the inspection plug on the trap.

It is a good plan to check the trap under the sink about once a year.

Bath and basin blockages normally respond to the plunger treatment.

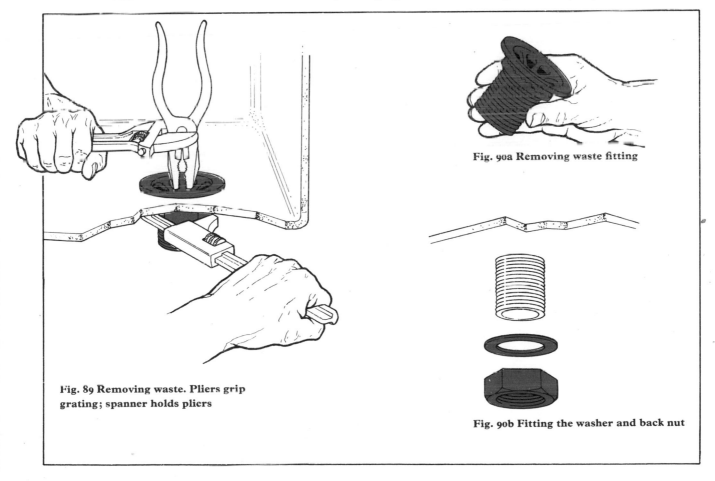

Fig. 89 Removing waste. Pliers grip grating; spanner holds pliers

Fig. 90a Removing waste fitting

Fig. 90b Fitting the washer and back nut

Fig. 92 Gullies should be cleaned out regularly

It is usually very difficult to get at a bath trap and the operation may involve removing the bath side panel. It is therefore worth persevering with the plunger treatment before going to this extreme.

SEALING A WASTE

Occasionally a waste fitting on a sink or bath will leak. To remedy this, first disconnect the trap underneath. Then get someone to hold the grating of the fitting steady, *fig. 89*, while you un-screw the back nut underneath with a wrench.

Take the waste fitting out of the outlet hole, *fig. 90a*, remove the old washer and clean up the fitting. Re-move all dirt and old putty and make sure you get the outlet thoroughly dry.

If the threads on the waste fitting are worn or damaged, renew the part. Apply some putty underneath the flange, put the fitting in the outlet and add putty to the washer. Replace the washer and the back nut, *fig. 90b*.

Get someone to hold the grating steady from above while you tighten up the back nut from underneath. Remove any surplus putty from the flange and underneath the back nut.

TOILET BLOCKAGE

The plunger treatment is sometimes also effective when there is a blockage in a w.c. pan. But if a block of toilet paper becomes wedged in the pan or the pipe it can be a difficult thing to shift. It may be possible, however, to dislodge it with a flexible cane.

In some houses, inspection plates are fitted at the soil pipe connection on an outside wall. These can be checked if a blockage of this sort cannot be cured by any other means.

A blockage in a w.c. on the ground floor may be due to an obstruction in the drainage system underground.

Fig. 93 A simple cover keeps rubbish out of gullies

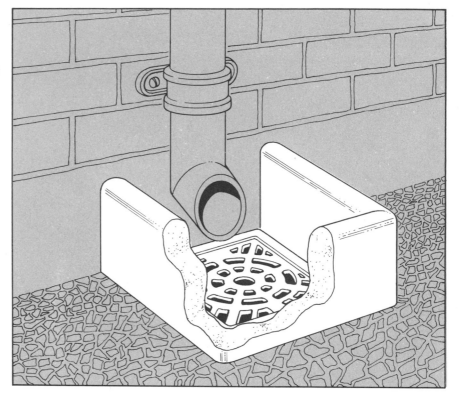

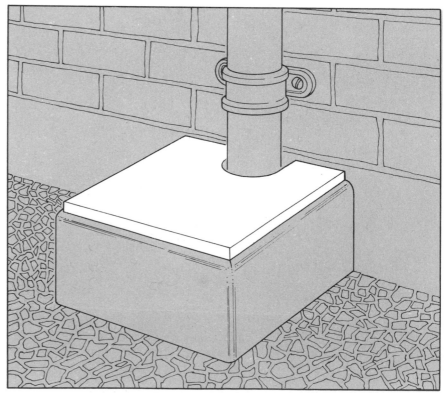

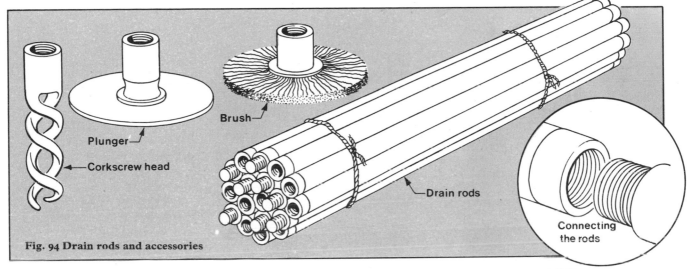

Fig. 94 Drain rods and accessories

This, of course, will involve an inspection of the manholes.

Gullies, *fig. 92*, are plumbing items which are often neglected. Consequently the grids can become choked up with leaves and other debris which gets blown into them. The result is that when soapy water is released from the sink, it does not flow to the gully and thence to the drains. So it overflows into the garden.

CLEAN THE GRIDS
The grids of gullies should be cleared of rubbish regularly and be cleaned and sterilised. This is a most unpleasant job but for the sake of hygiene it must be done.

Getting an old metal grid thoroughly cleaned and sterilised can be done by placing it in a fire for a few minutes. A better idea, though, is to replace it with a plastic type which is simple to keep clean.

There should, of course, be a concrete curb built around these gullies which helps to keep out a lot of the rubbish blown about in a garden. But it can't keep it all out, so the answer is to build a simple cover for it as in *fig. 93*.

UNDERGROUND BLOCKAGES
If water will not run away in a gully after the grid has been cleared of rubbish, the underground drainage system may be blocked up. This could also be the case if a ground floor w.c. pan fills nearly to the brim and is very slow to clear.

To find out where in the drainage system the fault lies, lift the cover off the inspection chambers or manholes and see where the water is lying. When you spot a dry chamber the blockage will be above that point (nearer the house).

To clear the obstruction, a set of drain rods will be needed. These can be screwed together like a chimney sweep's rods and can be fitted with various attachments, *fig. 94*. Some plant hire firms or builders' merchants hire them out.

A rubber plunger head is one attachment used for this purpose. Being flexible, it is capable of shifting soft obstructions.

Screw a few rods together and feed the plunger into the drain, twisting the rods in a *clockwise direction* as you do so, otherwise you may unscrew the rods and leave the leading rod and the plunger attachment stuck in the drain.

If the plunger fails to shift the obstruction, try a corkscrew head attachment but use it gently or you could fracture a pipe.

When finally the blockage is cleared, flush out the section thoroughly with a hosepipe while using the plunger disc to shift any remaining sediment.

FRACTURED PIPE
If, after all your efforts, the blockage cannot be cleared, the trouble may turn out to be a fractured pipe. An indication of its position can be obtained by counting the number of rods used in order to meet the obstruction.

This may well be under an impossible place to get at like a concrete path. In any case, all defects in the drainage system should be notified to the public health inspector. Repairs of this nature are normally best left to an expert. They should, in any case, be supervised by the health inspector's department, and tested and approved by him.

A knowledge of how a domestic drainage system works above ground, at least, is a valuable asset to any householder. There is much he can do to keep the system in working order.

TWO-PIPE SYSTEM
Take a look at *fig. 95*. If you can recognise your drainage system from the diagram, yours is what is known as a two-pipe installation. It may surprise you to learn that this system is now obsolete! However, don't despair. The majority of homes have similar installations.

A brief description is necessary. First there is the soil pipe to which the w.c. is directly connected, and which runs underground to empty into the sewer. Its other end should terminate at least 3 ft above the top windows; it acts as a vent pipe to allow the sewer gases to escape.

Every soil pipe should have a cage on its top to prevent it becoming blocked up with wind-blown rubbish and birds' nests.

Other pipes used in this system carry the rest of the house water to the sewer. These pipes discharge into open gullies.

HOPPER

In the typical system illustrated in *fig. 95*, both the bath and bathroom wash basin empty into a hopper on the wall outside. The hopper connects to a downpipe, which discharges into a gully at ground level.

In some systems the kitchen sink waste empties into the same gully via a short length of downpipe, but in other systems it has its own gully.

All the gullies mentioned are connected to the main house drain which leads to the public sewer.

In some old drainage installations both the house waste *and* the rainwater pipes are discharged into the same sewer, but in new buildings these are now normally separated.

In some even more primitive arrangements rainwater is disposed of via soakaways in the garden, but this practice is fast dying out.

DRAINS

Now to the drains themselves. Those which carry domestic waste to the sewer may be salt glazed earthenware or cast iron or, in modern buildings, pitch fibre or plastic.

Drains are laid in straight runs so that blockages can be cleared easily. However, if they have to change direction, or if a branch pipe joins the main drain, a manhole is provided at this point to permit access for clearing blockages, *fig. 96*.

In many areas the authorities lay down that the manhole should be provided near the boundary of the house at the point where the main drain falls to the sewer. To prevent gas from

the sewer building up in the house drains, this manhole is provided with an interceptor trap, *fig. 97*.

The rodding eye of this trap enables

any blockages which may occur between the manhole and the fall to the sewer to be cleared.

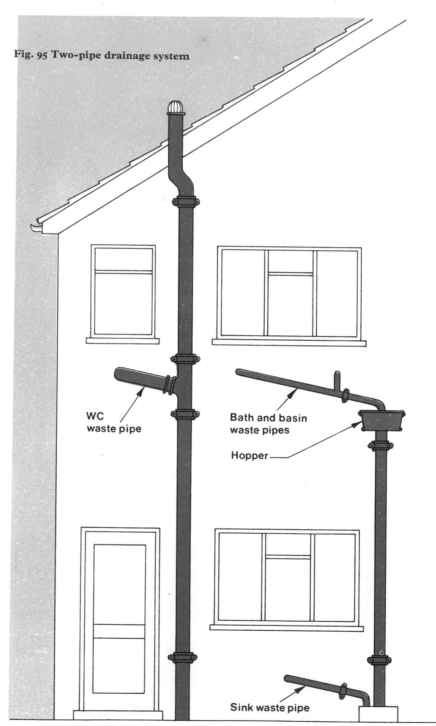

Fig. 95 Two-pipe drainage system

WC waste pipe

Bath and basin waste pipes

Hopper

Sink waste pipe

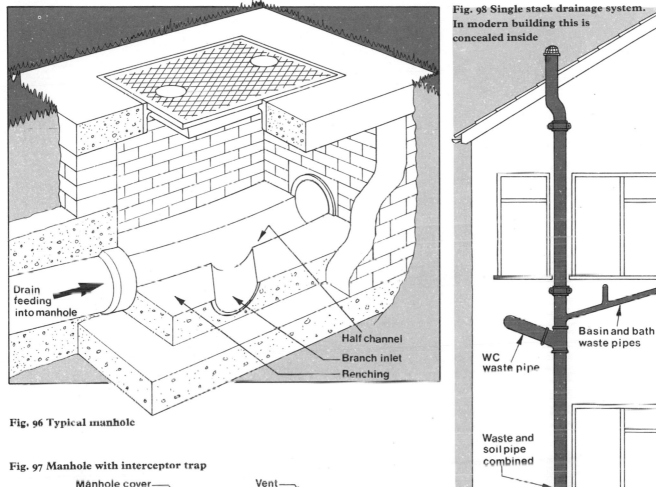

Fig. 96 Typical manhole

Labels on Fig. 96:
Drain feeding into manhole
Half channel
Branch inlet
Benching

Fig. 97 Manhole with interceptor trap

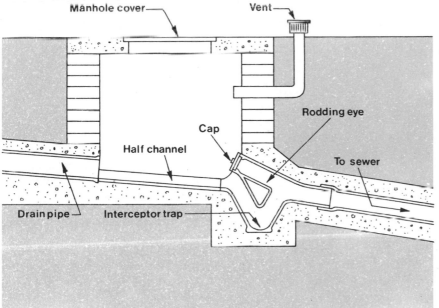

Labels on Fig. 97:
Manhole cover
Vent
Cap
Rodding eye
Half channel
To sewer
Drain pipe
Interceptor trap

Fig. 98 Single stack drainage system. In modern building this is concealed inside

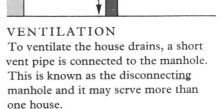

Labels on Fig. 98:
Basin and bath waste pipes
WC waste pipe
Waste and soil pipe combined
From sink
To drain (underground)

VENTILATION

To ventilate the house drains, a short vent pipe is connected to the manhole. This is known as the disconnecting manhole and it may serve more than one house.

The domestic waste is carried from the inlet to the outlet section of a manhole by half channels encased in benching. This benching is simply smooth finished concrete which assists the flow of waste water through to the sewer.

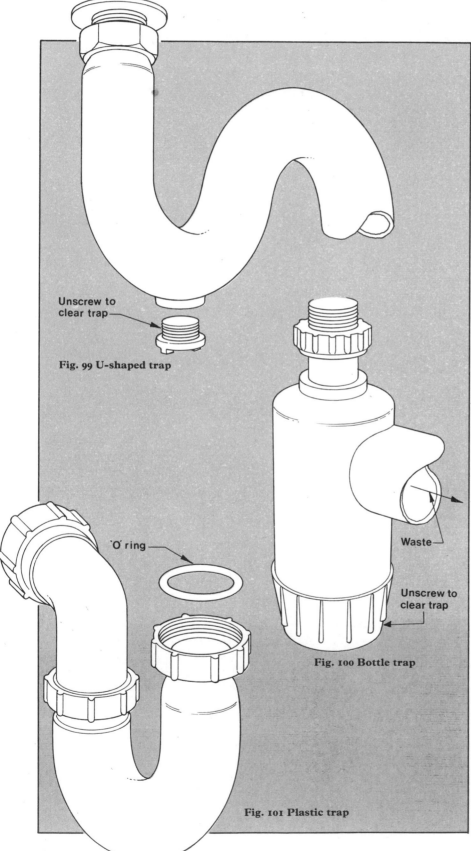

Unscrew to
clear trap

Fig. 99 U-shaped trap

'O' ring

Waste

Unscrew to
clear trap

Fig. 100 Bottle trap

Fig. 101 Plastic trap

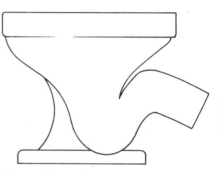

Fig. 102 S trap for ground floor w.c.

Fig. 103 P trap for upstairs w.c.

Manholes in some areas where there are modern developments have been replaced by a different method. This comprises a series of rodding points and involves the use of plastic drain pipes. Access to these points is provided by an inspection chamber or concrete cover.

SINGLE STACK

That is really all you need to know about your drainage system, but before leaving the subject, a word about drainage systems in new properties.

Under the Building Regulations, new houses are provided with what is called the single stack or one-pipe system, *fig. 98*. The difference in the new and the old systems will be appreciated if *figs. 95* and *98* are compared.

This new system is built inside the

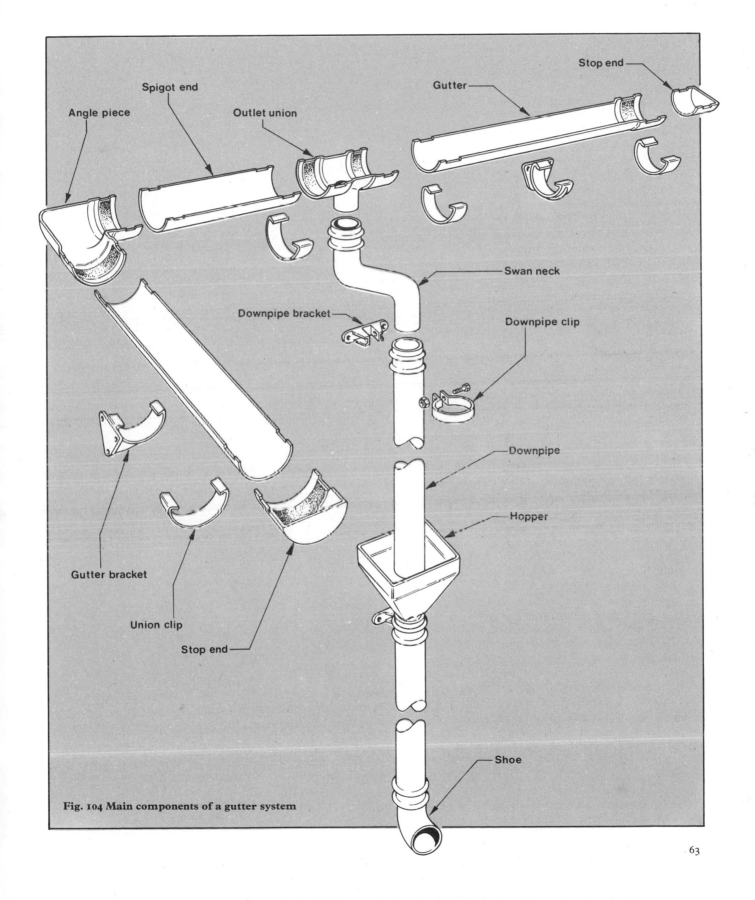

Fig. 104 Main components of a gutter system

house and all the wastes from sinks, basins, w.c.s and baths empty into the soil pipe (one-pipe). This pipe connects directly with the main drains of the house. It is vented above the level of the eaves and is served by the manholes already mentioned.

In some of these modern installations anti-siphonage vent pipes connect to the waste pipe of all the fittings (bath, sink, basin and w.c.) and to the soil pipe itself.

The purpose of the anti-siphonage pipe is to stop the water seal in a waste trap being sucked out by the force of the waste matter passing through the pipe from a fitting (perhaps a bathroom basin) on a floor above.

In some modern buildings, the waste traps of fittings are provided with deep seal traps as an alternative to the above system. In many of the modern single stack drainage systems, plastic is used instead of cast iron.

GULLIES AND TRAPS
We saw earlier how waste traps can give access to waste pipes in order to clear blockages. These traps are permanently filled with water to prevent odours rising from the drains and entering the waste pipes and permeating the atmosphere in the house by escaping through the w.c. pan and basin plug holes.

Many basin and sink traps are U shaped, *fig. 99*; others are bottle traps, *fig. 100*. In a plastic pipe system they may be shown in *fig. 101*.

In a ground floor w.c. it is usual to fit an S trap, *fig. 102*, but in an upstairs floor w.c., a P trap is usually installed, *fig. 103*.

NEW WASHER ON TRAP
Another simple plumbing job to be added to the d-i-y man's list is that of fitting a new washer on a leaking waste trap.

First put a bucket under the trap to catch the flow of water from it. Then unscrew the connecting nut as required. The leak may be at the top of the trap where it joins the sink, or at the bottom.

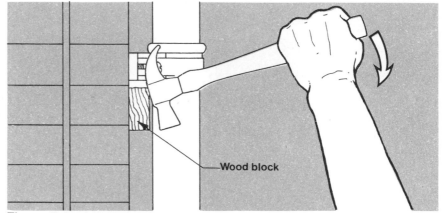

Fig. 105 Releasing downpipe bracket

Wood block

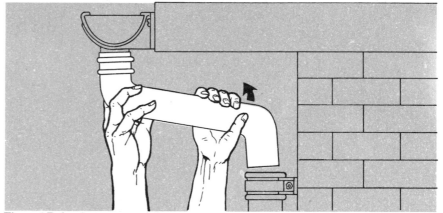

Fig. 106 Releasing a swan neck

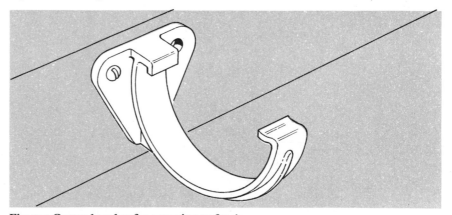

Fig. 107 Gutter bracket for screwing to fascia

Figs. 99 and *100* show two types of trap – the U shape and the bottle trap. After removing the faulty washer, take it with you as a pattern when you buy the new one. After renewing, turn on the tap and check that the leak has stopped.

GUTTERS
If he can work from a ladder, a d-i-y plumber should be able to maintain his gutters and downpipes in good working order.

Most gutters are now either cast-iron or pvc. While cast-iron types are extremely robust, they are liable to corrode if neglected. Therefore they should be examined regularly for signs of rust.

If a cast-iron gutter leaks at one or two points, this will probably be caused by corrosion which has eaten right through the material. Small holes can be repaired with an epoxy resin filler – Plastic Padding or Isopon, for example – after wire brushing the gutter to remove the rust (wear safety spectacles when you are doing this job or protect the eyes in some other way). But if a gutter is badly worn, it is better to replace the section affected by a new length.

Once corrosion starts in one length of gutter, it is quite likely that the remainder of the lengths will be similarly affected before long. If this is so, it is worth replacing the whole of the system with modern pvc materials.

Figure 104 gives an idea of how a gutter and its downpipes are constructed. Rainwater runs down the roof into the gutters which are fixed to fascia boards at the eaves. The gutters fall slightly to an outlet. This carries the water through the downpipe to the drainage system.

START AT THE TOP

There are various spots in the system which can become clogged up with leaves and other rubbish. So clearing them out should be made a regular task. Begin at the highest point and work downwards. Clean not only the gutters themselves, but all the outlets, swan necks, hopper heads and downpipes.

While doing this job, a lot of muck may pass through the downpipes into the gullies through the grids, so put a bowl or basin under the shoe of each downpipe to prevent the rubbish entering the drainage system. It is advisable to wear gloves on this job, especially if the gutters have sharp metal edges.

If the gutter is blocked, remove the debris but avoid letting it fall into outlets and thence to the downpipes.

CLEARING OBSTRUCTION

To clear an obstruction in a downpipe, tie a thick old rag securely to a cane and see if it can be pushed through the pipe. If this is not effective, you may have to disconnect the downpipes.

The pipes are fixed to the wall with brackets and clips. Use a claw hammer held against a block of wood to withdraw the nails from the bracket lugs, *fig. 105.*

The obstruction should be easy to remove once the length of downpipe is released. Flush the pipe through with a hose afterwards. Then refit the pipe. You will probably have to put new wooden wall plugs in the holes into which the nails can bite.

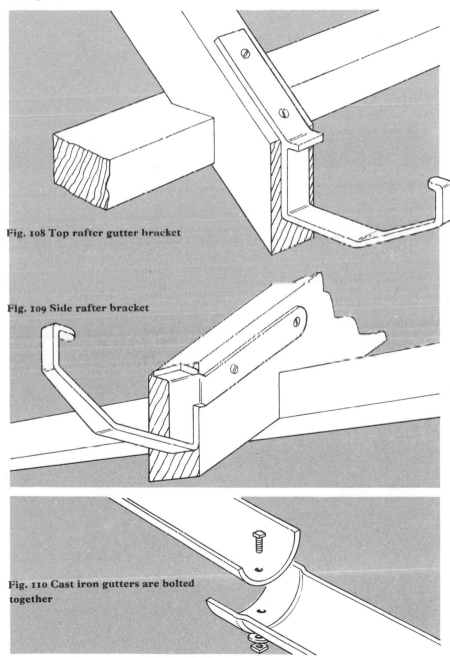

Fig. 108 Top rafter gutter bracket

Fig. 109 Side rafter bracket

Fig. 110 Cast iron gutters are bolted together

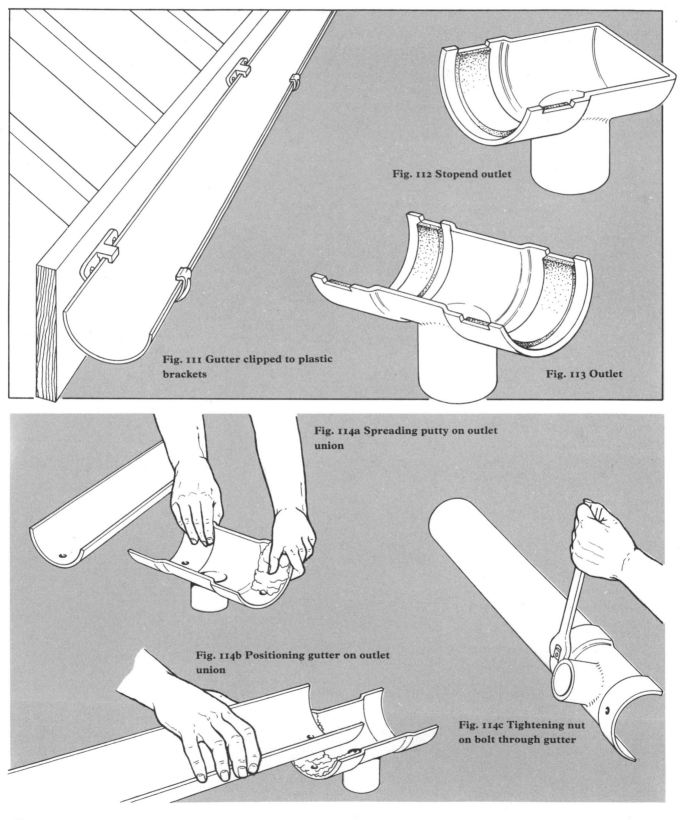

Fig. 112 Stopend outlet

Fig. 111 Gutter clipped to plastic brackets

Fig. 113 Outlet

Fig. 114a Spreading putty on outlet union

Fig. 114b Positioning gutter on outlet union

Fig. 114c Tightening nut on bolt through gutter

The joints of the pipes may be loose after taking them apart. If so, apply mastic to the joint and coat it with bituminous paint.

Swan necks, *fig. 104*, are also likely to become clogged with debris. To free a swan neck from a downpipe, push its lower part upwards to release it from the pipe. Then dismantle its other end from the gutter itself, *fig. 106*.

You will need two hands for this job so be careful you don't overbalance. Two pairs of hands, in fact, are better than one for jobs like this.

REMOVE RUST

Clean the section out thoroughly after dislodging the obstacle and remove any loose rust. Cast iron swan necks should be reconnected to the downpipe first, then to the gutter. If, however, the section is made of plastic, fix it in the reverse way.

When you have removed all the debris, tip some water in the gutter at its highest point and see if it runs away quickly. If the gutter has sagged at any point, water will tend to build up there and may cause the gutter to overflow.

Faults of this nature should receive immediate attention or the wall will become soaked with water and, in a solid wall especially, may cause internal damp problems.

SAGGING GUTTERS

There are two ways to cure a slight sag in a gutter. You can try raising the length of gutter concerned by fixing a new gutter bracket to support it where it sags, or the section between it and the downpipe can be lowered but not by more than I in. This movement is restricted to prevent the joints between the gutter sections being forced apart.

There are two main types of gutter bracket, *figs. 107* and *108*. If yours are rafter brackets, you will have to remove a slate or tile to get at the fixing rafters. The other type screws to the fascia.

If the brackets are cast-iron types, they will probably be bolted to the gutters, *fig. 110*. Plastic brackets are simply clipped over the gutters, *fig. 111*.

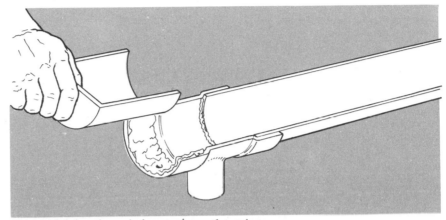

Fig. 115 Joining the end piece to the outlet union

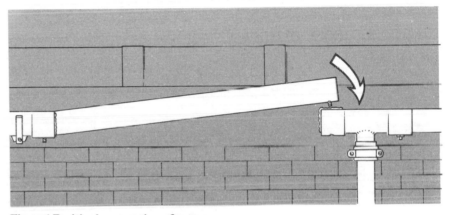

Fig. 116 Positioning a section of gutter

NEW LENGTH OF GUTTER

To replace a length of cast-iron gutter, first clean out any debris and remove the swan neck or, if there isn't one, the downpipe. Next, the nut holding the gutter to the bracket should be undone, but this will most probably need to be oiled first to help release it. If it won't budge, you may have to cut right through the nut with a small hacksaw and renew it.

When the section has been released from its brackets, lift it carefully and lower it to the ground. It will be heavy and brittle (if old), so don't try to go it alone!

Cut the new gutter, if necessary, to length with a hacksaw, using the old length as a guide. If there are any joints, allow for overlaps. Drill a $\frac{5}{16}$ in (8 mm) diameter hole for the bolt.

The outlet union (which may be a stopend or intermediate type), *figs. 112* and *113*, will come away with the section of gutter. Clean this up (or renew if necessary) and spread a layer of metal casement putty on the inside edge, *fig. 114a*. Press the new length of gutter into the putty – aligning the holes on both gutter and union, *fig. 114b*.

Press a bolt into the hole, add a washer and nut and tighten up, *fig. 114c*. Clean off all excess putty.

Cut off a new end piece and drill bolt holes as before. Line the stopend (new or old) with putty, bolt it to the end piece and join to the outlet union, *fig. 115*.

PAINT IT

Before fixing the new section of gutter, paint it inside with bituminous paint. Renew the gutter brackets if necessary,

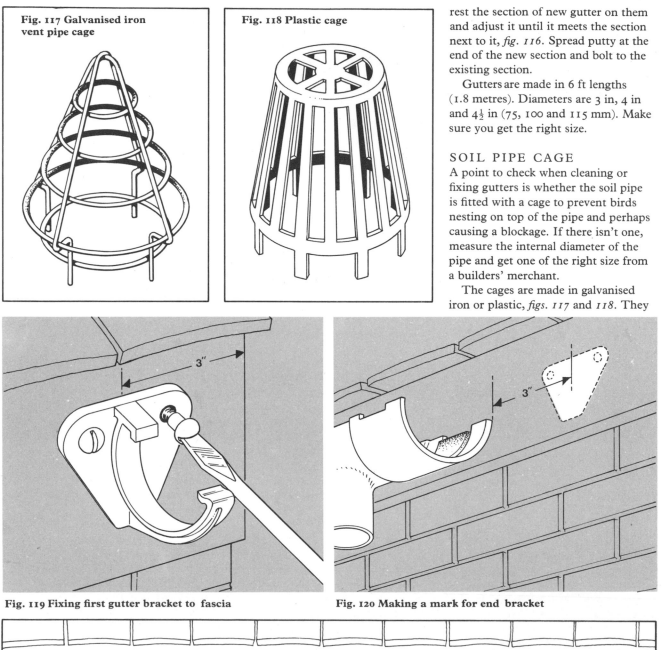

Fig. 117 Galvanised iron vent pipe cage

Fig. 118 Plastic cage

rest the section of new gutter on them and adjust it until it meets the section next to it, *fig. 116*. Spread putty at the end of the new section and bolt to the existing section.

Gutters are made in 6 ft lengths (1.8 metres). Diameters are 3 in, 4 in and 4½ in (75, 100 and 115 mm). Make sure you get the right size.

SOIL PIPE CAGE
A point to check when cleaning or fixing gutters is whether the soil pipe is fitted with a cage to prevent birds nesting on top of the pipe and perhaps causing a blockage. If there isn't one, measure the internal diameter of the pipe and get one of the right size from a builders' merchant.

The cages are made in galvanised iron or plastic, *figs. 117* and *118*. They

Fig. 119 Fixing first gutter bracket to fascia

Fig. 120 Making a mark for end bracket

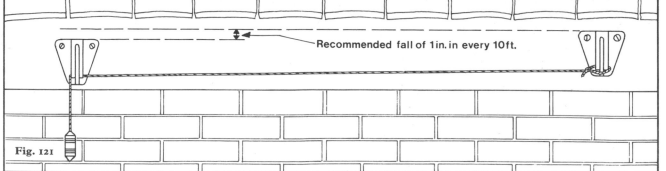

Recommended fall of 1 in. in every 10 ft.

Fig. 121

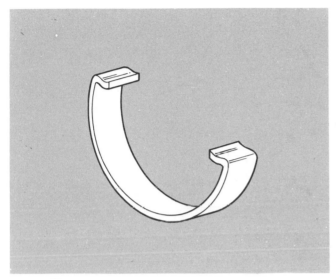

Fig. 122 Gutter strap

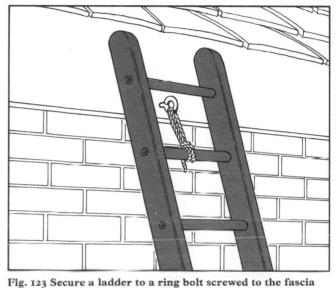

Fig. 123 Secure a ladder to a ring bolt screwed to the fascia

are easily fitted by squeezing the prongs and pushing the cage into the top of the pipe as far as possible.

PLASTIC RAINWATER FITTINGS

There are several manufacturers of plastic rainwater fittings. If you plan to replace your gutter system with plastic materials, it is a good idea to get pamphlets from a dealer or from the manufacturers themselves. These include Marley and Osma Ltd. (Marley Plumbing, Dickley Lane, Lenham, Maidstone, Kent. Osma Plastics Ltd., Rigby Lane, Dawley Road, Hayes, Middx.)

You will find instructions and illustrations on how to fit the components and these should be followed implicitly.

A gutter should slope (fall) gradually from its closed end down to the outlet pipe (downpipe). The recommended fall is 1 in in 50 ft.

Methods of fixing plastic gutters vary according to make. Here is a brief outline of the procedure but watch for variations in the pamphlet supplied. The pamphlet should also give directions on welding and jointing where this is applicable.

Having taken down the old gutter,

clean the fascia board and paint it if necessary. When the paint is dry, fix your first bracket 3 in from the top end (the one at the opposite end to the downpipe, *fig. 119*). At the other end, hold the stopend outlet where it is to be positioned and make a pencil mark 3 in from it for the final bracket, *fig. 120*.

STRING LINE

Run a string line attached to the bracket at the other end and tie a weight to it such as a plumb bob. Hold the bracket against the fascia and let the string line hang over it, *fig. 121*. Check that the line is horizontal by using a spirit level. Then fix the bracket slightly lower so the water will flow towards the downpipe.

Fix the rest of the brackets in position at the distances recommended. Make sure all the joints have brackets near them for support. Fix the outlet and the gutter lengths, linking them with jointing pieces.

There may be variations in the types of jointing pieces, but generally they are gutter straps, *fig. 122*, which clip on to hold the spigot and socket of the gutter in a watertight joint.

The lengths of downpipe are fitted from the top downwards and you add

the necessary supports according to the manufacturer's instructions.

SCAFFOLDING IS SAFER

Fixing gutters really needs two pairs of hands, as I said earlier, and it is much easier to work from some form of scaffolding rather than from ladders. If you have to use a ladder, make sure it is anchored firmly by tying it to a ring bolt screwed into the fascia board, *fig. 123*.

Don't lean a ladder against a plastic gutter. Being smooth surfaced, a ladder can slip on these gutters very easily. For another thing, plastic gutters are flexible, which means that they are not really suitable as ladder supports.

Plastic gutters should not be painted. In fact, they are most difficult to paint because the release agent used in the mould to shape the gutters adheres to their surface for some time afterwards and it is not easy to get paint to 'take' on their surfaces.

If, however, you want to paint them so that they match the rest of the exterior decorations, don't do so for at least a year. This will give the release agent time to be washed off by the rain. Then paint them with two coats of gloss paint. Do *not* use primer or undercoat.

Outside Taps

One tap in the house which does a lion's share of the work is the cold water tap at the kitchen sink. Even when the garden needs watering or the car requires hosing down, it is frequently this tap which is used, frequently at an inconvenient time for the housewife!

Fitting an outside tap is the obvious solution and it is a job within the scope of a handyman, provided he has at least some idea of how a pipe should be cut and can take accurate measurements.

My advice is to make certain you know exactly what is involved before you attempt this job because it will mean cutting the rising main pipe. If a hash is made of this part of the operation, then you will be in trouble!

When the job is being done, the domestic water supply will have to be turned off, so make sure your cold water storage cistern is full and, as a safeguard, first draw off some drinking water from the kitchen tap for cooking and other purposes.

TURN OFF THE WATER
Let's assume that your rising main pipe is either copper or stainless steel and that it emerges from the floor or the wall under the kitchen sink. A few inches above the floor there should be a stop-valve and draincock (they may be combined) fitted to the rising main pipe. If this is so, you can turn the water off at this point. Otherwise use the stopcock buried in the pavement outside.

The rising main pipe above the stop-valve may then go direct to the kitchen tap, or there may be a tee-jointed branch pipe, fig. 124.

A new branch pipe to your outside tap should be taken from a point *above* the existing stop-valve, but *below* the branch pipe supplying the kitchen tap, fig. 125.

You may therefore be restricted in your choice of position for the branch to the outside tap. But remember not to fit the branch too low down. Allow plenty of room for buckets or watering cans to be held under the tap outside.

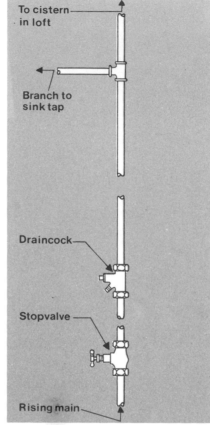

Fig. 124 Branch to sink tap

The branch pipe to the garden should slope *slightly* downwards so that it can be drained completely in frosty weather.

MATERIALS
Measure up carefully for the amount of pipe needed. Allow a little extra as a precaution and for any bends to be made. I will assume that compression joints will be used as these are simpler for an amateur who is working in a confined space.

Apart from ½ in (15 mm) copper pipe, you will need

One tee connector (½ in or 15 mm), fig. 125.

One stop-valve (may be referred to at the shop as a stopcock) 15 mm × 15 mm, fig. 125.

One wall plate elbow with compression joint inlet and threaded female outlet (to accept tap), fig. 126.

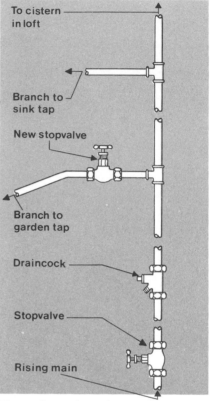

Fig. 125 Insert branch to garden tap below branch to sink tap but above the stopvalve

One hose union bib tap.

I suggest you buy all the fittings at the same place, explaining what they are needed for. You can then try them out and see that everything fits neatly together. If possible, I suggest you buy a bib tap which will incline forwards. This prevents grazing your hands on the wall when operating the tap.

If you are pushed for time, it is better to do the job in two parts. The first thing to do after cutting off the water supply and draining the rising main pipe through the draincock is to fit the tee connector.

CUT THE PIPE
Hold the connector against the rising main pipe and carefully mark off the section of the pipe to be cut out to make way for the tee, ie. the distance between the shoulders on the inside

of the connector. Be very careful not to cut away too much pipe.

As you will be working in a confined space, it may be easier to cut the pipe with a small hacksaw rather than with a tube cutter (as described on pages 43 and 44). Cut the pipe squarely and trim off all the burrs.

Remove the nuts and olives (rings) from the tee connector and slip them over the cut ends of the pipe. One end of the pipe may have to be sprung slightly to get it into the connector. Tighten up the nuts over the olives and the tee joint is almost complete.

You will now need to cut off a short length of copper pipe from your new length and join it to the branch outlet (the remaining hole) of the connector. Tighten up as before.

FIT THE STOP-VALVE
The new stop-valve is now fitted to the other end of the short length of pipe in the same way. Take care to ensure that the arrow on the stop-valve is pointing *away* from the rising main but *towards* the garden tap. If the valve is fitted incorrectly, no water will be able to flow through to the tap outside!

Now you can turn this stop-valve off. This completes the first part of the job. The water can now be turned on again to bring the house supply back to normal. Water cannot now flow through the new stop-valve, and the rest of the operation can be completed at any time.

The next thing to do is drill a hole through the brick wall for the pipe to the tap. For this you will need a long masonry drill to be fitted in a power drill. If you have a solid wall, this will be about 9 in thick. A cavity wall, of course, will be about 11 in thick, including the cavity.

You need a hole big enough to take a ½ in (15 mm) pipe. The job *can* be done with a cold chisel and hammer, but it is not a method I would choose.

If the pipe has to change direction to enter the hole in the wall, you can use an elbow bend compression fitting, to avoid bending the pipe. This is the easiest way to do the job.

Cut off another short length of pipe and fit one end of it into the vacant end of the stop-valve and its other end into the compression bend. Now cut off enough pipe to run from the com-

pression bend to the wall plate elbow outside.

BEND THE PIPE
Note that this pipe will have to be bent slightly *downwards* towards the garden tap. You can, of course, use another compression bend to achieve this change of direction, but it is less expensive to bend the pipe using a ½ in bending spring.

Grease the spring, then drop it into the pipe to the spot where the pipe is to be bent. Bend the pipe over your knee. As you do this, the spring inside the pipe will act as a support and prevent the pipe from becoming flattened.

When the bend is complete, insert a metal rod into the ring on the bending spring to act as a tommy bar. Turn the spring clockwise and pull it out of the pipe.

Now push the pipe into the hole in the wall and connect it to the compression bend inside or, if no bend is needed inside, direct to the stop-valve.

WALL PLATE ELBOW
The position of the wall plate elbow

Fig. 126 **Wall plate elbow**

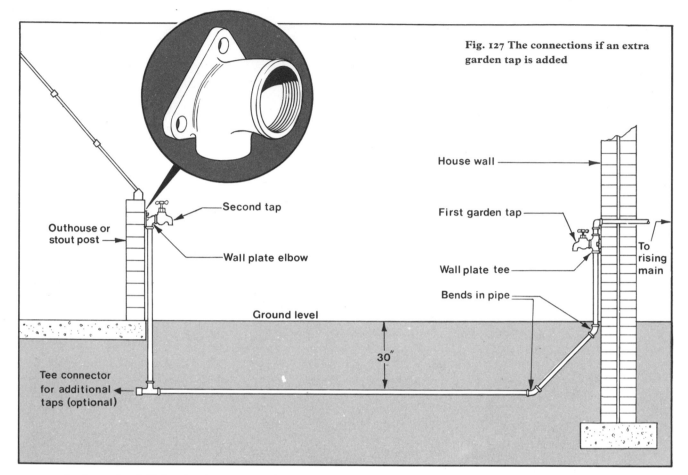

Fig. 127 The connections if an extra garden tap is added

Outhouse or stout post

Second tap

Wall plate elbow

Tee connector for additional taps (optional)

Ground level

House wall

First garden tap

Wall plate tee

Bends in pipe

To rising main

30"

can now be determined. Holes should be drilled in the wall and plugged for the fixing screws. Now join the pipe to the elbow, *fig. 127.*

To make sure of a watertight joint, bind ptfe plastic thread sealing tape around the thread of the tap before screwing it into the wall plate elbow.

Your garden tap is now complete and you can turn on the new stop-valve. If the outside tap should drip or is hard to turn, you may find that some bits of brass filings have found their way inside. This is common in some new taps.

If so, remove the headgear of the tap, as described on page 34, and see if any filings are in the bottom of the body or elsewhere.

They may prevent the water being turned off completely. Unscrew the tap from the wall plate and clean it out as necessary.

Before replacing the headgear in the tap's body, apply Vaseline to the thread. This will make it easier to remove if and when it needs a new washer.

In the winter the water supply to the garden tap should be turned off at the new stop-valve and the garden tap drained as a frost precaution.

EXTENSION TAP
Having got a water supply freely available on an outside wall you may want another tap fitted near the top of the garden. This is now no problem. If you decide to do this, instead of fitting a wall plate elbow to the wall as described, fit a wall plate tee. The pipe to the second tap can be joined to the bottom outlet.

It is best to use polythene tube instead of copper pipe to supply the second tap, and it should be buried

about 30 in below ground level to protect it from gardening tools thrust into the soil.

Similar compression joints to those used for the first tap will be suitable for polythene tube. You can get special inserts which are designed to stop the tube collapsing as the nuts on the compression joints are tightened.

With polythene tube, you usually have to get compression fittings a size larger than those used to join copper pipes. So if you use $\frac{1}{2}$ in (15 mm) polythene tube you will require $\frac{3}{4}$ in (22 mm) fittings.

ADAPTORS
You can get polythene adaptors which will greatly simplify the process of changing over from copper to polythene.

These have a plain end which fits into an existing compression joint. The

Fig. 128 A patio swimming pool
installation in the author's garden

other end is an enlarged joint to take
the polythene tube. The fitting is
joined to the bottom of a wall plate tee
and your polythene tube is connected
to it.

When you have made the connection
at the tee, bury the length of polythene
tube in a trench 30 in deep. Near its
other end fix a timber post in the
ground to which the second tap can be
attached. Use hardwood for the post
and protect it against the weather by
treating with preservative.

Screw a wall plate elbow to the post
and your second tap to the elbow,
fig. 127.

You will have to bend the polythene
tube in two places – where it is first
buried and where it rises to join the
second tap.

To bend the tube, you will need
water kept at boiling point. Immerse
for a few minutes and bend as required.
Once the tube has cooled off, the bend
should remain.

Remember to drain the pipe and
turn off the water supply when the cold
weather arrives. You won't get the
water out of the pipe which is underground, of course, but as the pipe is
polythene, no harm can arise even if
the water in it should freeze up.

Scale and Corrosion

One great disadvantage of a direct hot water system (see also chapter 3) is that it is more prone to scale problems than an indirect system – especially in hard water areas.

When scale forms, the flow pipe becomes furred up near the boiler. This restricts the circulation of the water. If steps are not taken to prevent this happening, back circulation will eventually take place.

The hot water will force its way back up the return pipe, *fig. 129*. This will cause all sorts of gurgling and banging noises which will be especially noticeable when the boiler fire is burning well.

When a system produces these noises, it can be dangerous and should be descaled as soon as possible. Before the noises are heard, it may be noticed that the boiler will be taking longer to fill the cylinder with heated water.

UNWANTED INSULATION
When scale forms in a hot water system it acts as an insulator and a type of insulator you can well do without! When scale forms inside a boiler it actually insulates the water against the heat produced by the boiler fire. This is why the water in the cylinder takes longer to heat up.

The remedy for this is *not* to stoke up the boiler to increase the heat. By doing this, you will cause the system to overheat and thus produce even more scale. Then the noises will increase in volume as the water tries to force its way through the pipes. Overheating can eventually cause the boiler to leak.

Fortunately, it is possible for a d-i-y man to descale his own hot water installation by introducing chemicals which will remove the layers of scale (the insulators). You can buy descaling kits just as you can the small packets of descaler which are designed to remove scale from kettles.

Albright & Wilson Ltd., PO Box 3, Oldbury, Warley, Worcs., is one firm which offers descaling kits. Drop them a line if you can't get hold of a kit locally. Full instructions are supplied.

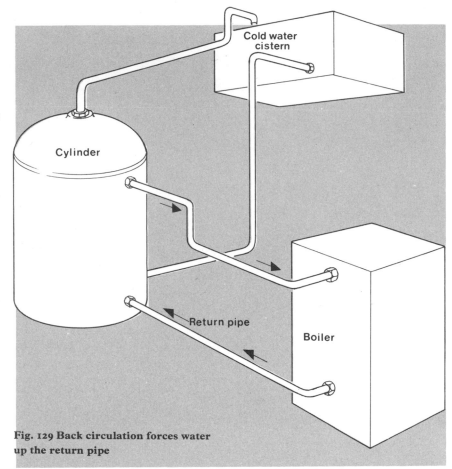

Fig. 129 Back circulation forces water up the return pipe

The chemicals are put into the system via the cold water storage cistern or through the draincock on the boiler.

PRECAUTIONS
There are one or two things you can do to avoid the system becoming scaled up to the extent described. First and foremost, don't let the water boil in the boiler. Although many older types of boiler which use solid fuel cannot be temperature controlled, in time you get to know the 'feel' of the thing and can judge when the water is getting near boiling point.

When in doubt, you can do the obvious thing – run some of the hot water off before it boils. This may be wasteful, but it is kinder to your system in the long run.

If you have a boiler fitted with a temperature control, there is no prob-

lem. The temperature of the water has to reach about 160° Fahrenheit before scale starts to form to any appreciable extent, so if the temperature can be kept at no higher than 140°, the risk of scale is considerably reduced. The thermostat on the immersion heater should also be set at 140° for the same reason.

ADDITIVES
Another thing you can do is to introduce chemical additives into the water. These will have the effect of stabilising the chemicals which are responsible for scale forming.

A well-known one is Micromet, made by Albright & Wilson Ltd. This is a simple device consisting of a polythene mesh basket, *fig. 130*. This is suspended in the cold water storage cistern. Although it prevents scale in the *hot* water system, it must be hung

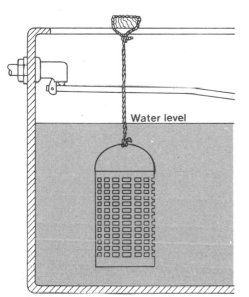

Fig. 130 Micromet crystals in container to prevent scale

in the *cold* water cistern which supplies that system.

The basket is submerged in the water near the ball-valve inlet, but clear of both the ball float arm and the bottom of the cistern.

The life of the Micromet is six months. The makers will send you a reminder every six months that a refill is due.

This device does not soften water and it is not intended for use in closed-circuit central heating systems.

Another problem can arise in *soft* water areas. Iron and galvanised parts of a system may corrode causing tap water to become rusty and leave stains on the sanitary ware. You can use Micromet for this purpose also. It controls this type of corrosion and eliminates rust-coloured water.

INDIRECT SYSTEM
The most drastic cure for scale is to replace a direct hot water system with an indirect one, which would be less likely to scale up.

The reason for this is that a given volume of water contains only a certain quantity of the chemicals which can cause scale to form. Therefore, once these have been deposited in the primary circuit of an indirect system

(when it is first fitted) no more scale can form. The water is used over and over again, and only small losses which are caused by evaporation have to be replaced.

The secondary circuit of an indirect system (the one which supplies the domestic hot water) will not normally become hot enough to cause much scale to form.

CORROSION
Electrolytic corrosion can occur in galvanised steel cold water cisterns and hot water tanks if copper and galvanised steel are both used in the same plumbing installations. We don't need to worry too much how it occurs, but briefly the cause is this.

When, say, the steel of a cistern is galvanised to protect it against corrosion, it is coated with zinc. Some types of water are charged with a large

amount of carbon dioxide which can act as an electrolyte (a sort of electric conductor).

An electric current flows between the zinc on the cistern and the copper pipes. This causes the zinc to dissolve, leaving the steel exposed. Rust will then form on the cistern.

PAINT THE CISTERN
This can be prevented by painting the inside of a *cold* water storage cistern with bituminous paint. This paint must be tasteless and have no smell. (This method cannot be used on a galvanised steel hot water tank.) The paint will stop the water coming into contact with the galvanising on the cistern and causing corrosion.

Have a look at the cold water cistern up in the loft, or wherever it is. If you find traces of rust inside it, but the outer sides of the cistern are in good

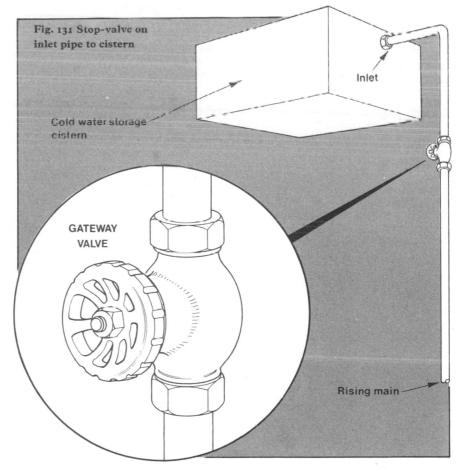

Fig. 131 Stop-valve on inlet pipe to cistern

Cold water storage cistern

Inlet

GATEWAY VALVE

Rising main

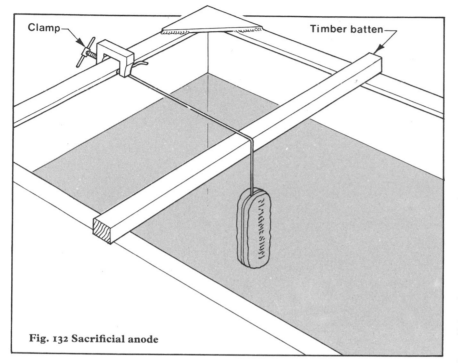

Clamp

Timber batten

Fig. 132 Sacrificial anode

order, then you can extend its life by coating it with this type of paint.

First you will have to shut off the water supply to the cistern, drain and wait until it is thoroughly dry. You may be able to speed up the drying process by using a hair drier.

Tie up the ball-valve to make sure no water can seep past the stop-valve.

REMOVE THE RUST
The first job is to get all the rust off by wire brushing (protect your eyes) and by using abrasive paper. As you remove the rust you may find that the surface of the cistern is pitted in places. These areas should all be filled in with a resin filler (Isopon or Plastic Padding, for example).

Then give the cistern two good coats of the bituminous paint, making sure the first is dry before the second is applied. This is obviously a job which cannot be done in a few minutes. Most of the plumbing system will be out of action while it is being done.

However, if there is a stop-valve fitted on the inlet pipe (the rising main) to the cistern, fig. 131, it will not mean that the whole of the cold water supply

to the house will have to be cut off. Water will still be available from the cold water tap at the kitchen sink.

Treating a cistern as described should add a few more years to its life, but if the rust is too widespread and the cistern shows signs of leaking, then it should be scrapped and replaced, preferably with a type that cannot rust.

SACRIFICIAL ANODES
There is a good way of preventing corrosion in a plumbing system. This involves fitting a sacrificial magnesium anode in the hot water tank and cold water cistern.

This is simply a lump of magnesium which has a higher electrical potential than zinc. When it is fitted into a cistern or tank it must make good electrical contact with the metal.

The electrolytic action will then occur between the zinc on the cistern and the anode (which is made of metal) itself. This will cause the anode to dissolve, or be sacrificed, thus protecting the galvanising on the cistern.

You can buy these sacrificial anodes easily. A well-known one is called the Mapel Tank-Saver and there is a

different type for cold water cisterns and for hot water tanks.

The cold water type is hung in the water and the copper wire attached to it is fixed to the edge of the cistern by clamping it, fig. 132. First, though, the metal part of the cistern where the wire is to be fixed must be scraped clean to ensure perfect electrical contact.

HOT WATER TYPE
Fitting a Tank-Saver to a hot water tank, fig. 133, is slightly more complicated. To begin with, first drain the tank from the draincock next to the boiler. Then remove the cover of the tank.

Drill a $\frac{1}{4}$ in diameter hole in the centre of the cover and screw the Tank-Saver in place. Make sure of a good electrical contact by scraping the metal thoroughly around the hole.

There are other types of anode available, including the Metalife Solid which is simply stood in the base of either a hot water tank or cold water cistern.

You should be able to buy or order these anodes through a plumbers' or builders' merchant.

MAKING REPAIRS
Repairs can be made to galvanised steel cisterns and tanks and copper cylinders using glass fibre matting and polyester resin repair kits. The tank, cistern or cylinder must be drained first, of course, and thoroughly dried. All rust must be removed as already described. Take great care that all the rust dust is removed – preferably by a vacuum cleaner.

Cold water cisterns can be repaired from the inside; hot water cylinders from the outside. To mend a hole in a cold water cistern, cover the hole from the outside with a piece of card taped on to the cistern so that the repair material cannot fall through the hole, fig. 134. The inner side of the card can be covered with wax to prevent the resin adhering to it, or waxed paper can be inserted between the card and the cistern.

Three pieces of glass fibre matting are now needed, each one big enough to overlap the area under repair by about 1 in all round.

After mixing the hardener with the resin, according to the manufacturer's directions, apply a coat of it about $\frac{1}{8}$ in thick to the area and bed a piece of matting into it.

Apply another coat of resin and add a second layer of matting. Repeat the operation and leave the repair to harden. Use the same process to make a repair to a hot water cylinder from the outside.

Read the instructions in your kit carefully and if they vary from the method outlined, follow them implicitly.

There are many other repair jobs which can be tackled around the house using one of these kits. In the plumbing field, these include burst or leaking pipes, cracked radiators, chipped metal baths and cracked basins.

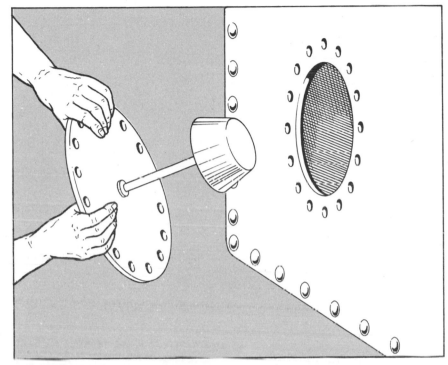

Fig. 133 Tank-Saver fitted to cover of hot water tank

Fig. 134 The stages of corrosion in a tank

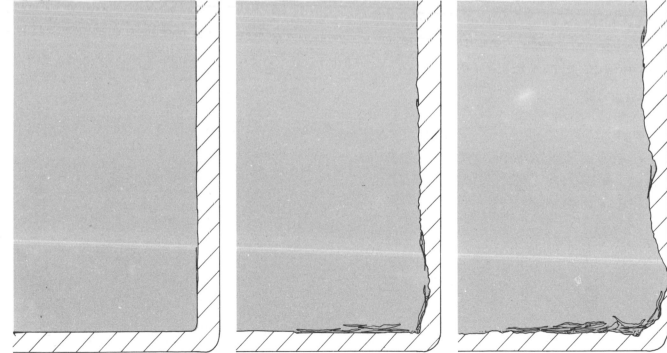

Condensation

Two rooms in a house which suffer more from condensation than any other are the bathroom and kitchen. This is due to the high volume of water vapour produced by cooking, washing up, bathing, washing and so on. What can be done to reduce the condensation?

The answer is to increase ventilation, provide steady controlled warmth, insulate thoroughly and do everything you can to cut down the volume of water vapour produced.

It sounds very simple but unfortunately it isn't as easy as it sounds. There are many ifs and buts.

Both the bathroom and kitchen have cold surfaces wherever you look. In the bathroom there is the bath, the basin, possibly exposed pipes, a w.c. cistern, probably wall tiles and glass. In the kitchen we have the sink, plastic laminate worktops, possibly exposed pipes, washing machine, fridge and various other appliances, plus no doubt partly tiled walls.

When warm air which is filled with moisture hits these surfaces when they are cold, it immediately cools down below what is known as its dew point. The water the air contained then turns into droplets rather like early morning dew on the grass.

WARMTH

Steady warmth of the correct kind will help to ensure that all the cold surfaces I have mentioned, and others besides, will not cool off below dew point.

One of the best forms of heating for this purpose is an electric convector or fan heater, or an oil-filled radiator. Don't use portable oil or gas heaters which are not fitted with flues as they generate vapour and will cause even more condensation, excellent though they may be for other heating purposes. A heated electric towel rail in a bathroom will also help to keep condensation down.

Heating alone, however, is by no means the complete answer to the problem. Some ways have to be found to make sure that whatever moisture-laden air there is in the room can

Fig. 135 Extractor fans ensure proper ventilation

Fig. 136 Fit a cooker hood to dispel cooking odours and vapour

Fig. 137 Another type of extractor fan

escape and be replaced by drier air from other parts of the house.

AIR EXTRACTORS

The best way to ensure that is to fit an electric air extractor, *fig. 135*, in the window of the room concerned. You need a type which can be switched on when moisture is being produced.

Fitting a cooker hood, *fig. 136*, which is ventilated to the air outside, will dispel the vapour caused by a cooker.

There is not a great deal that can be done to reduce the amount of vapour produced by normal domestic activities. As you may well argue, no-one produces steam or vapour deliberately! There are small things, however, that may be overlooked which would undoubtedly help to keep the condensation at least at a tolerable level.

BATHROOM CONDENSATION

In the bathroom, for example, do you run several gallons of hot water into the bath before adding cold water? If you do, this will cause a heavy volume of vapour (steam, if you prefer) to form, so try the reverse method. Run a couple of inches of cold water

first, then turn on the hot tap and note the difference in the amount of steam which rises from the bath. This method also helps to protect the bath surface.

If, for any reason, you are unable to fit an electric towel rail in the bathroom, get a radiant-type electric heater fixed really high up on the wall. It must be out of reach of anyone in contact with the bath and be operated by a pull cord switch.

If you can turn the heater on some time – say half an hour – before running the bath water, this will take the chill off the cold surfaces in the room and reduce condensation.

I have said elsewhere that the temperature of 140° Fahrenheit is adequate for domestic hot water. If the temperature can be kept at this level, less condensation will be produced when the hot water is run off at any point.

INSULATION

Insulating the house thoroughly will also be a great help in keeping condensation at bay. Although the subject is too complex to be covered adequately in this book, I would emphasise the importance of insulation for this purpose.

Apart from reducing the amount of heat lost through walls, ceilings, windows and so on, adequate insulation of the areas immediately above both the kitchen and the bathroom, and the addition of expanded foam polystyrene tiles on the ceilings of these rooms, will do much to cut down the formation of vapour.

Cavity infill for cavity walls is one of the most effective ways of insulating the whole house, but unfortunately this

is not a job you can do yourself. You need a contractor armed with the right equipment to inject the special insulation materials.

Solid walls are more of a problem and cannot, of course, be treated in this way, but covering them with sheet polystyrene before hanging your wallcoverings will keep the walls warmer and prevent a lot of the condensation which might otherwise form on them.

And while talking of wallcoverings, a vinyl type has a higher insulation value than ordinary wallpaper (even washable types) and are much easier to strip, by the way, when redecorating.

DOUBLE GLAZING

It has often been claimed that double glazing will cure condensation, but this is far from true in most cases. Much depends on the type of double glazing installed. If factory sealed units of a reliable make are used, no condensation should form between the outer and the inner panes of glass. But it can form on the inside of inner panes in some cases if the room gets cold.

With the cheaper forms of double glazing there is no guarantee that condensation will not appear on the insides of the panes if the unit is not sealed, and if any moist air in the room can get to them.

Properly installed, however, any efficiently sealed double glazing unit will *reduce* condensation on the glass and will, of course, cut the amount of heat lost through the panes. But a tubular heater or radiator fixed underneath a window will probably be more effective in preventing the glass from becoming steamed up.

Showers

No bathroom can be described as fashionable or really up to date if it does not contain a shower. No longer is a shower considered a luxury. Indeed, in these days when economy is one of our main concerns, a shower is a most useful appliance which has several advantages over the traditional bath.

Apart from the fact that a shower is hygienic and convenient, it uses considerably less water than a bath – a fact which, perhaps, is not generally recognised. A shower is quicker to take than a bath and the space needed to accommodate it is far less.

There are types of shower small enough to be fitted into an area 2 ft 6 in square. It will be seen, therefore, that a shower need not necessarily be installed in a bathroom upstairs.

Given certain essentials, which I'll deal with in a moment, a shower can be fitted in a spare room, in a corner of another room (perhaps a bedroom), or even in a hall or on a landing. It would normally be no problem to disguise its presence by erecting a simple enclosure for it – a challenge, in fact, to a handyman's ingenuity!

WATER PRESSURE

An essential requirement for a shower is sufficient water pressure or head at the outlet, *fig. 138*. If this pressure is not adequate, the shower will be a disappointment.

To get an adequate head, the base of the cold water storage cistern, which will normally provide the water supply, should be at least 3 ft above the level of the shower outlet, and 4 ft or 5 ft would be better still.

It is not important at what level the *hot* water cylinder or tank is positioned. It can be higher, lower or on the same level.

Ideally, too, the cold water side of the shower should be fed by a separate pipe from the cistern. The reason for this is that no water from it can then be drawn off the pipe. If there are other draw-off points, the pressure to the shower will fall when these are used and the temperature of the water will

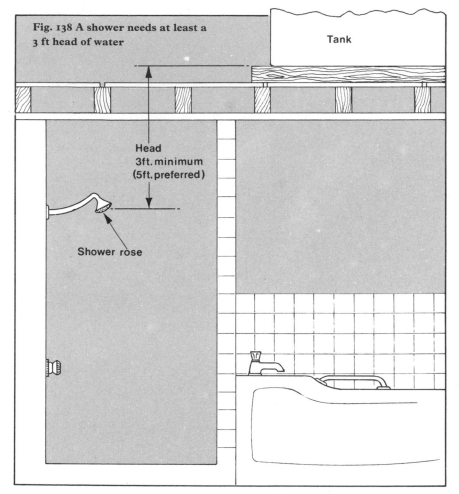

Fig. 138 A shower needs at least a 3 ft head of water

Tank

Head
3ft. minimum
(5ft. preferred)

Shower rose

increase suddenly. There will then be a danger of the occupant of the shower being scalded!

The same remarks apply to the hot water supply. If water can be drawn off this supply pipe from elsewhere, the hot water pressure will fall and the occupant will receive a freezing cold shower.

Regulations do not normally allow cold water supplies direct from the main to be mixed with hot water supplied by a gravity system. The cold water supply should come from the cold water storage cistern.

SPECIAL FITTINGS

There is, however, one way in which this problem can sometimes be overcome. There are special fittings available in which the hot and cold water

supplies are channelled separately to the point where the water is discharged and are mixed just below the sprinkler. So long as the hot water pressure is sufficient, a shower can be fitted in this way and it will not contravene the regulations.

If the cistern is not high enough to ensure adequate pressure, it may be possible to fix it at a higher level by building a raised platform for it. If you decide to do this, however, bear in mind that the platform will have to carry a great weight!

To do this will involve lengthening the pipes leading to and from the cistern, but this can be done by using simple compression fittings. The alternative way to increase the water pressure is to fit a pump, but this will increase the cost considerably.

WATER HEATERS

If it is not possible to run a shower off your cold water system, there is an alternative method which may be acceptable. This involves fitting a gas or electric instantaneous water heater.

The heater has to be a special approved type of appliance. Not all heaters are satisfactory for this purpose so you should make careful inquiries, not only from the shower supplier, but from the local gas or electricity board as well.

These appliances must also be acceptable to local water authorities and bylaws governing their acceptance or otherwise may vary from area to area.

So make careful inquiries before launching into the purchase of this type of equipment.

These heaters are connected direct to the main, the heater control valve and sprinkler combining in one unit.

MIXING VALVES

The mixture of hot and cold water needed for a shower can be obtained by fitting a mixing valve. This will equalise the pressures of the hot and cold water supplies.

Mixing fittings can be separated into two categories. The first is simply a pair of hot and cold taps whose outlets are joined together by a mixing chamber. The required temperature is reached by adjusting the taps.

The other is a mixing valve. This combines the two taps in one and is operated by a single control. With this type, the required temperature is chosen manually by moving a pointer round a scale which is usually marked hot, warm, tepid and cold.

THERMOSTATIC VALVES

Some mixing valves are thermostatic. Showers fitted with a thermostat can adjust the water temperature automatically when it runs too hot or cold. This is the type of valve which should be used if separate supply pipes are not run to the shower from the cistern and cylinder.

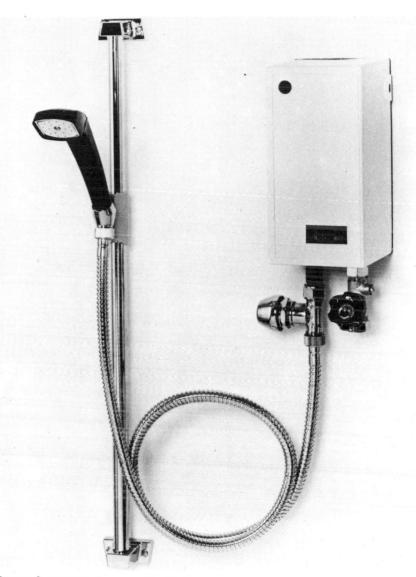

Fig. 139 A compact shower unit with its own water heater

The non-thermostatic type is a screw-down valve and has an indicator for 'hot', 'tepid', 'cold' and 'shut'. These valves are operated by a single control which turns through cold before the tepid and hot settings are reached. This is a British Standards stipulation. The best models of this type of valve have washers, and sometimes seatings, which can be renewed.

SHOWER HEADS

There are two types of shower head (the final outlet for the water) available. One is referred to as a rose, fig. *144*, like that on a watering can, even though some models may not look exactly like one. However, they all work on the same principle, and the water is expelled through a number of small holes.

The rose type of shower head will normally operate from a slightly lower head of water than the other type – the sprayer.

The sprayer types are fitted with somewhat larger holes to discharge the

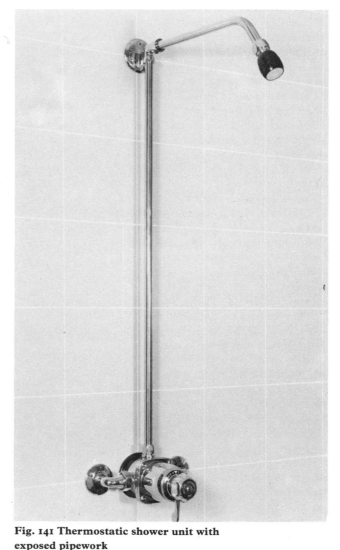

**Fig. 140 Pipes are concealed with this
type of shower fitting**

**Fig. 141 Thermostatic shower unit with
exposed pipework**

water, but are more economical in the
amount of water they use. The larger
holes, of course, make the sprayer
easier to clean, and it is less likely to
become clogged up than a rose type.

On some models the size and volume
of the spray produced can be varied by
changing the end cap. With some types
of shower head the pipes can be concealed, *fig. 149.*

TYPES OF SHOWER

A shower can be fitted in some circumstances so as to be used in conjunction
with a bath. In other cases, the bath is
dispensed with and the shower occupies

part of its space; or the shower can be
fitted as a separate unit, space per-
mitting of course. The method chosen
will obviously depend on the amount of
room available.

If the shower is to be separate from
the bath, or is to replace it, the first
requirement is a shower tray, *fig. 145.*
These trays are made in a number of
materials including enamelled iron,
glass fibre and stainless steel. Some are
finished with ceramic tiles.

The shower cubicle can be pur-
chased complete with plinth, tray, a
curtain for the open side, shower head
and mixing valve. These cubicles can

be positioned more or less anywhere in
the house where connections for the
hot and cold water supply pipes and
the waste can be conveniently made.

If you choose a cabinet of this type,
get one with a top provided, as this will
prevent the vapour escaping and
causing condensation in the room.

An alternative is to build the shower
in a corner using the existing walls for
two of its sides. An angled tube can
then be fitted to both walls to carry a
curtain wide enough to be draped along
its full length. You can get plastic
shower curtains in many colours and
patterns to match the decor of the room.

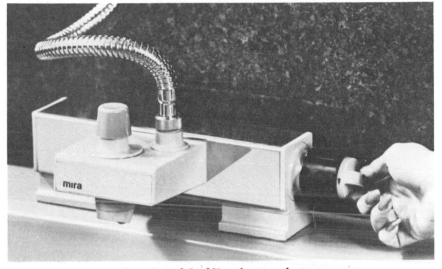

Fig. 142 Bath-mounted version of the Mira shower selector

If a shower is to be fitted above the bath, and the bath used for standing in instead of a tray, make sure that the curtain is long enough to fall inside the bath to prevent splashing. The usual place to fit the shower is at the end where the bath taps are positioned.

KEEP PIPE RUNS SHORT

When installing a shower of any sort, keep the length of pipe runs to a minimum. This will prevent a big build up of hot water in the pipes which will cool rapidly while the shower is not being used.

Not a great deal of plumbing work is involved in fitting a shower. If one is to be fixed over the bath, a thermo-

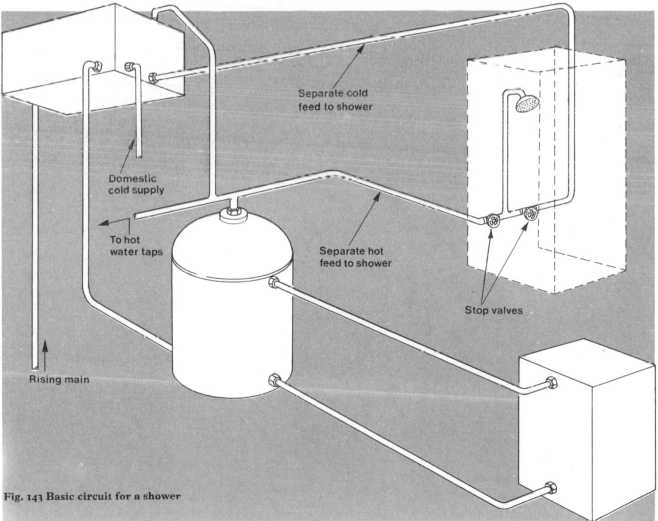

Separate cold feed to shower

Domestic cold supply

To hot water taps

Separate hot feed to shower

Stop valves

Rising main

Fig. 143 Basic circuit for a shower

Fig. 144 Rose-type shower head

Fig. 145 Shower tray

Fig. 146 Simple but strong wooden framework for supporting a bath

Fig. 147 The wall plate for the shower mixer is screwed to the wall

static mixing valve can be combined with the bath taps.

If the shower is to be separate, $\frac{1}{2}$ in copper or stainless steel supply pipes can be used and connected to the mixing valve. Non-manipulative compression joints can be used to join the the pipes and fittings.

When fixing the supply pipes to the mixing valve, make sure the joints are watertight. Wrap ptfe plastic thread sealing tape around the male threads of the fittings before screwing the connections up tight.

Figure 143 shows the basic connections for a simple shower.

The waste disposal can prove a problem in some circumstances. If, however, the shower is to be fitted in a bathroom, or other room on the ground floor, the waste pipe can pass through a hole made in the wall and discharge into the most convenient gully.

Where a shower is to be fitted in an upstairs bathroom in an old house, the waste pipe can be taken through the wall and discharge into the hopper head.

As shown on page 61, however, many homes are now fitted with a single stack drainage system. In these houses, all the waste from bath, bathroom basin, w.c. and sink is discharged into a single waste pipe. So if you fit a shower, its waste will also have to be discharged into this pipe.

Here you should consult the local council health inspector before trying to make the connections as it may be contrary to the local bylaws.

If your shower is to be independent of the bath you will need a $1\frac{1}{2}$ in waste and a P trap. Some shower trays are fitted with an overflow outlet; others don't have one.

In the latter type, no stopper should be fitted to the waste outlet. To fix the waste, pass it through the outlet hole provided in the tray and bed its flange in a mastic jointing compound.

Underneath the tray, push some more jointing compound up between the threads of the waste and the underside of the outlet hole, put the washer on and screw up the back nut. Then fit the P trap to the waste and the waste pipe to the trap.

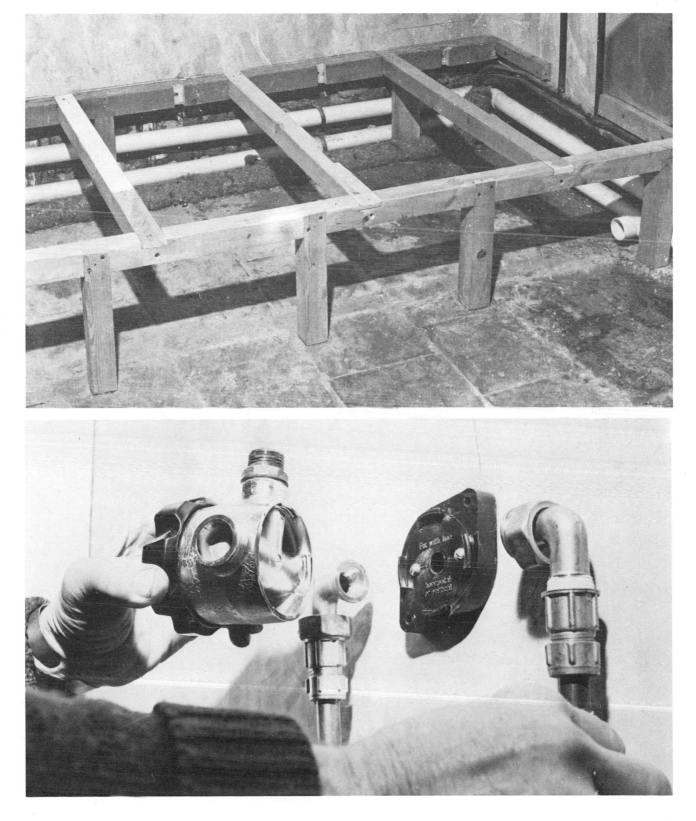

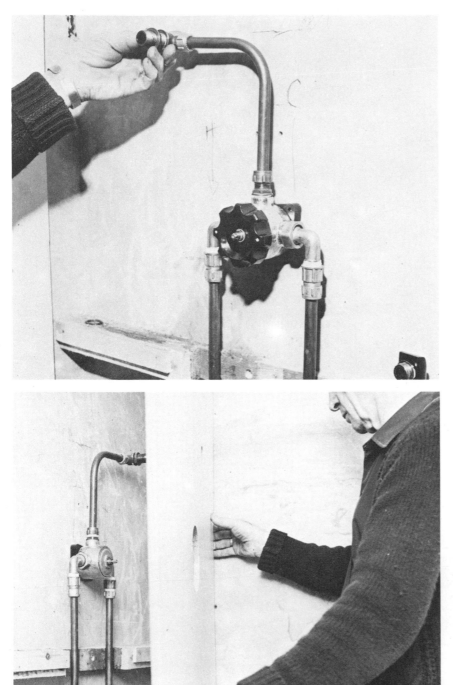

Fig. 148 The mixer pipe has a right angle connector for the flexible cord attachment

Fig. 149 Concealing the pipework with a waterproof panel

Fig. 150 The concealing plate is set in the hole over the mixing valve

Fig. 151 The black flow control is slotted over the spindle, followed by the temperature control

Fig. 152 Marking out the position of the hand spray mounting

Fig. 153 The finished shower installation

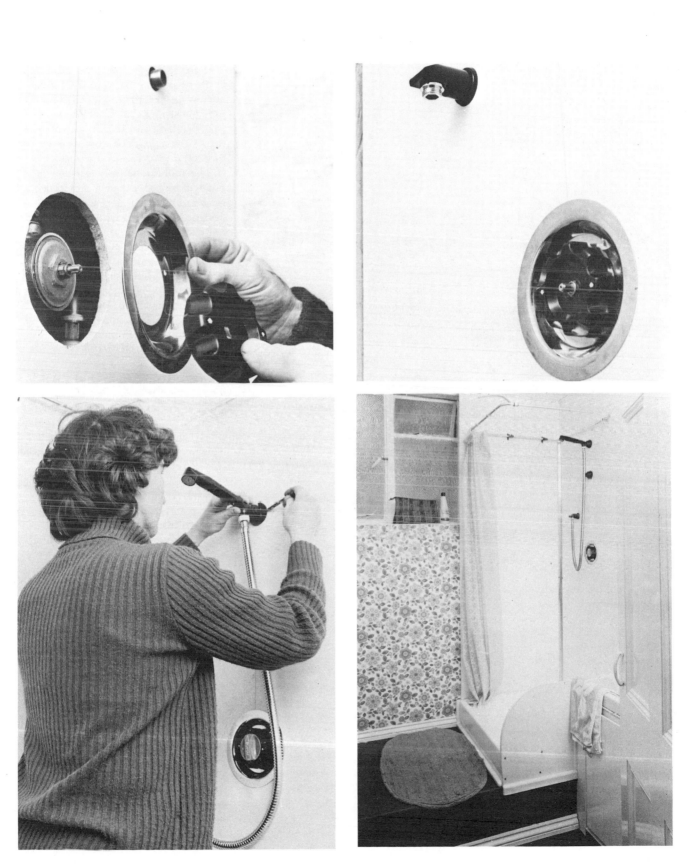

Fig. 154 Plumbing connections to a bath

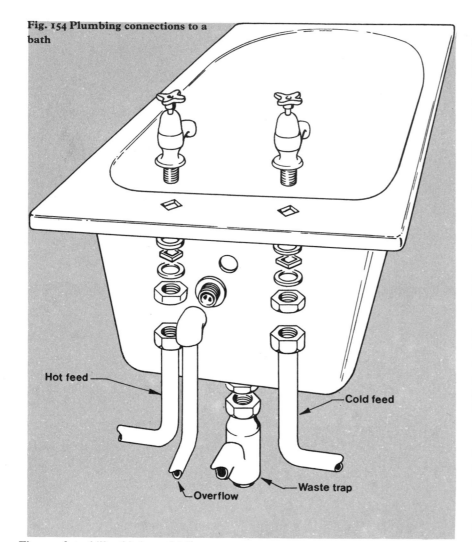

Hot feed

Cold feed

Overflow

Waste trap

Fig. 155 A tool like this is useful for working in confined spaces

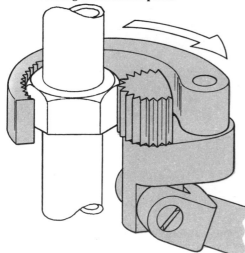

If you think you can cope, make sure at the beginning that you have, or can borrow or hire, the proper tools for the job. Those nuts will probably take considerable effort to shift, so you will need a strong bath or a basin crow's foot spanner or wrench *fig. 156*.

This tool can be used vertically (as you will probably have to use it) or horizontally. There is one size which is designed for coping with the taps' back nut and another, larger, version for dealing with wastes and traps. You will need both unless it can be found by prior experimenting that an ordinary wrench will do both jobs.

After removing the bath panel and the frame to which it is fitted, you will be able to see the plumbing connections.

DISCONNECT THE TRAP
The first job is to unscrew the nut which connects the trap to the outlet of the bath, *fig. 156*. The overflow may also be attached to this trap and will have to be disconnected.

You will now have to turn off the water before going any further. If the hot and cold water to the bath is supplied from the cold water storage cistern, there will be no need to turn off the main stopcock on the rising main. Simply turn off the gate valves or tie up the ball-valve. This will ensure that the supply of cold water to the kitchen sink will not be interrupted.

In a well-designed hot water system, there should be no need to put out the boiler fire. Better still, though, tackle the job in the summer months when the boiler is out. The light will be better for working in at this time of the year in any case.

Next, turn on the hot and cold water taps at the bathroom basin and the hot tap at the kitchen sink. This will drain the cold water storage cistern and the supply pipes.

DISCONNECT THE TAPS
Now undo the nuts which connect the taps to the supply pipes, *fig. 157*. This

Dismantling an old bath and installing a new one would appear to be a formidable task to many people. Certainly this is by no means a simple job, and one that calls for at least some practice in unscrewing awkward nuts which may be badly corroded and in a confined space.

It is therefore advisable to think seriously before tackling the job. Take a long, cool look at your existing bath. Remove the panel and see whether access to the taps and waste is going to be feasible. The chances are that you will have to work in inadequate light while lying on your side or, at best, on your back. *Figure 154* shows a typical bath plumbing arrangement.

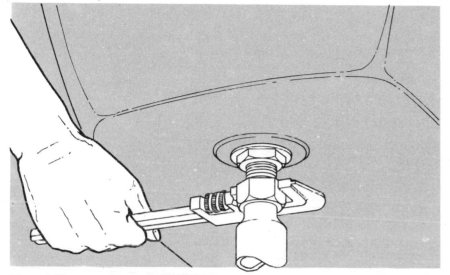

Fig. 156 Disconnecting bath waste trap

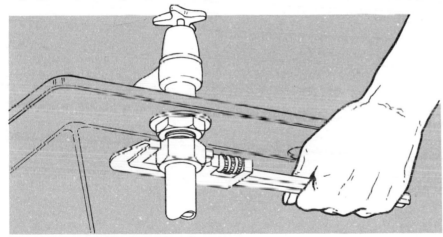

Fig. 157 Disconnecting bath pipes from taps

is easier said than done and you may need to use penetrating oil first.

The bath can then be removed. You will need help at this stage as the old type of bath is very heavy and cumbersome. The taps can now be removed from the bath.

If there is going to be a delay before the new bath is fitted, the taps can be reconnected temporarily to the supply pipes and the water turned on again.

Cast iron baths can be broken up in the bathroom with a club hammer, but it is advisable to smother the bath with a blanket or something similar to prevent chunks of the thing flying around and causing damage – possibly to you!

FIXING THE NEW BATH
Methods of fixing a new bath vary according to make. If you have chosen one of the modern plastic baths, this will be supplied with a specially-made cradle to support it. This should be fitted according to the accompanying instructions.

Enamelled steel or cast iron baths usually have four lugs which are screwed to the underside. The feet are then screwed into the lugs and the back nuts tightened up. To do this, you will have to roll the bath on its side and put a block or other support under it so you can get at the lugs comfortably.

When the feet have been attached and the bath lifted into position, you

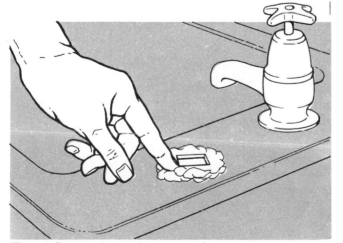

Fig. 158 Set taps in mastic compound

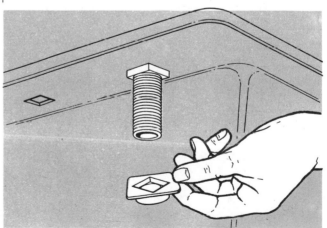

Fig. 159 Adding a top hat washer to the tail of a bath tap

can fit the taps. These can be set in a bed of mastic sealing compound, *fig. 158*. If you plan to use the old taps again, it is a good idea to first examine their washers and renew them if necessary.

Put a thick rubber or polythene washer (known as a top hat) on the thread of the tap before adding and

tightening up the back nuts, *fig. 159*. Before fitting the pipes to the tails of the taps, wrap some ptfe tape around the threads to make sure the joint will be watertight. If the tails of the new taps are shorter than the old ones, you can get a small adaptor which will lengthen them by $\frac{1}{2}$ in.

Now fit the bath waste, *fig. 160a*,

bedding it into a mastic sealing compound and again insert a thick washer between the bath and the back nuts, *fig. 160b*.

That completes the operation. But before you fit the bath panel, turn the water supply on again and release the tied-up ball float arm in the cistern. Turn on both the bath taps and check that the connections are not leaking and that the water runs away down the waste.

ADD THE PANEL
When buying your new bath, you probably ordered a bath panel to go with it. In case you didn't, and for those who want to make a panel for their existing bath, here's one way to do it, *fig. 161*.

You will need a timber frame to which the panel can be fixed. Battens of 2 in × 1 in timber will be satisfactory. One should be fitted firmly to the floor and another parallel with it will run along the top. Fit upright battens at each end and spacing battens about

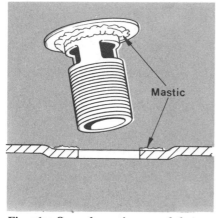

Fig. 160a Spread mastic around the waste outlet

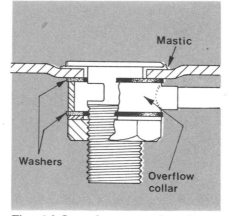

Fig. 160b Spread more mastic under the waste outlet; add rubber washers

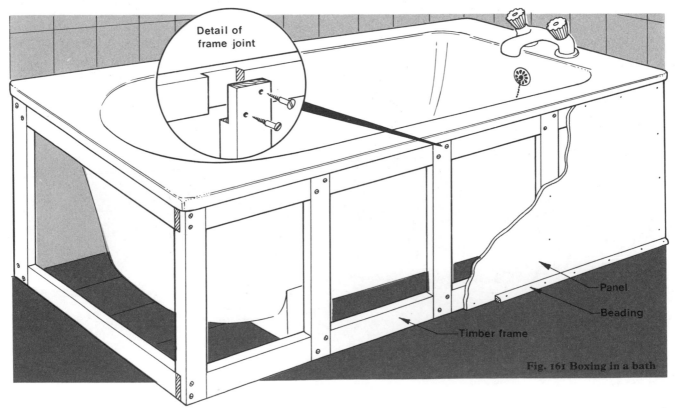

Fig. 161 Boxing in a bath

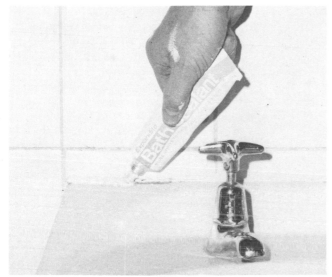

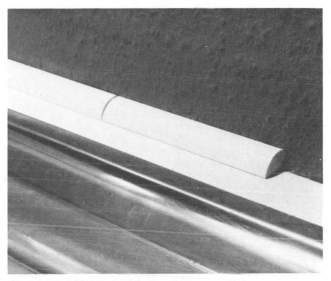

Fig. 162 A rubber-based sealer can be used to fill the gap between bath and wall

Fig. 163 Quadrant ceramic tiles can also be used to fill the gap

1 ft apart between them.

This frame will need anchoring to the wall and how this is done will depend on where the walls are.

The frame can then be covered with ordinary hardboard and then painted, or you may prefer to fit plastic-faced hardboard. The facing materials should be fixed with rustless screws and their heads covered with plastic domes.

Alternatively, ordinary hardboard can be covered with flexible plastic sheeting such as Con-Tact or Fablon. Either will provide an easy-to-clean surface which can be renewed when required.

FILLING THE GAP

After you have fitted your new bath, you may find that a gap may gradually develop between it and the wall tiles. There are a couple of ways you can fill this gap.

The first method is to fill it with a rubber-based sealer which is flexible to allow for any movement of the bath. Make sure that the gap is free of dirt and grease and, of course, thoroughly dry.

Squeeze the sealer into the gap with a steady movement, working forwards, *fig. 162*. If any parts of the sealer are uneven, they can be levelled with a

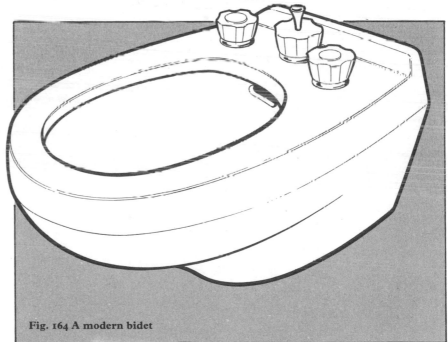

Fig. 164 A modern bidet

moistened finger. Any uneven edges can be trimmed with a knife or razor blade.

Another way to seal the gap is to fix quadrant tiles along it, *fig. 163*. These are normally supplied in packs and instructions are included. Make sure you use the adhesive recommended by the manufacturer.

BIDETS

The bidet is far more popular in this country than it used to be and can be regarded as a beneficial accessory to health and hygiene.

Some local authorities have special bylaws concerning the installation of a bidet, so these should be checked first before any arrangements are made to

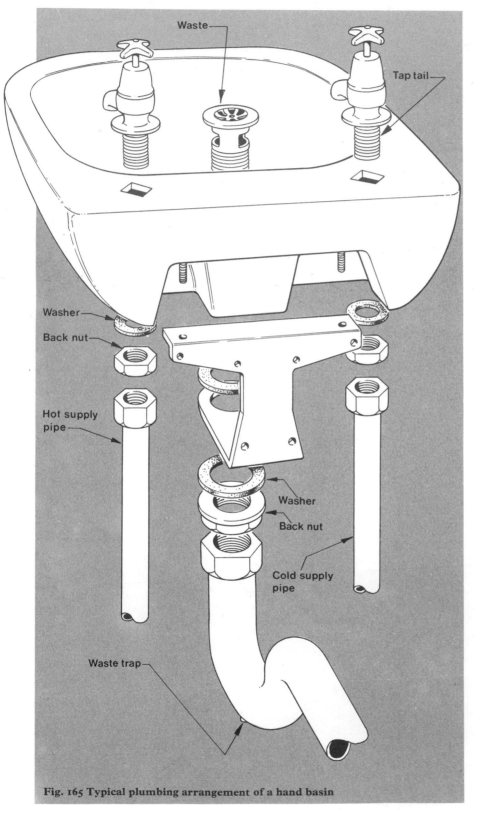

Waste

Tap tail

Washer

Back nut

Hot supply
pipe

Washer

Back nut

Cold supply
pipe

Waste trap

Fig. 165 Typical plumbing arrangement of a hand basin

have one fitted.

Some bidets are 'all rim supplied':
the rim seat is warmed by hot water
which fills the bowl for washing. Alternatively, the supply of water can be
diverted through a douche which may
have a spray or jet nozzle.

Other types of bidet have an 'over
rim' supply. These are simpler types
which have no douche but simply the
bowl for washing purposes. This is
supplied in the same way as an ordinary
washbasin.

To avoid back siphonage, most water
boards have special regulations for the
fitting of bidets with douche facilities.
The regulations mentioned below do
not apply to bidets with an over rim
supply.

Generally speaking, requirements
are (a) that bidets are supplied with
cold water from the cistern and not
from the main, (b) both hot and cold
water supplies should be drawn direct
from the source and there should be no
branch pipes teed off to any other
fitting.

WASHBASINS
If you are competent enough to fit a
new bath, the job of fixing a new washbasin will not daunt you. One point to
watch is that many washbasins are
supplied with safety brackets for fitting.
These should be used whenever possible to ensure stability. Do not rely on
the plumbing connections alone for
support.

Before fitting a wall basin, make sure
the wall has sufficient load bearing
strength to support it.

Another important point is that
when fitting the waste into the basin,
the slot in the overflow must line up
with the opening which is incorporated
in the basin, *fig. 165*.

The pipework will be similar in most
respects to that for a bath. Arrangements of pipes vary, but a likely
installation is shown in *fig. 165*.

If you choose a pedestal-type basin,
the pedestal will hide much of the
pipework. A good arrangement is to
run the hot and cold water supply pipes

up through the bathroom floor. The waste pipe should, if possible, pass through the wall behind the basin, but much will depend on the situation.

The taps can be fixed with Fix-a-Tap sets described earlier and compression joints can be used for the pipe connections.

Some basins are supplied without brackets, their weight being supported partly by the pedestal and partly by the pipes.

To fix the basin, put it on its pedestal and mark the position on the wall. Prop up the basin securely, remove the pedestal and connect the pipes. Slide the pedestal underneath and bed the basin to the pedestal with Plumber's Mait compound.

Figs. 166 and 167 Two modern bath designs

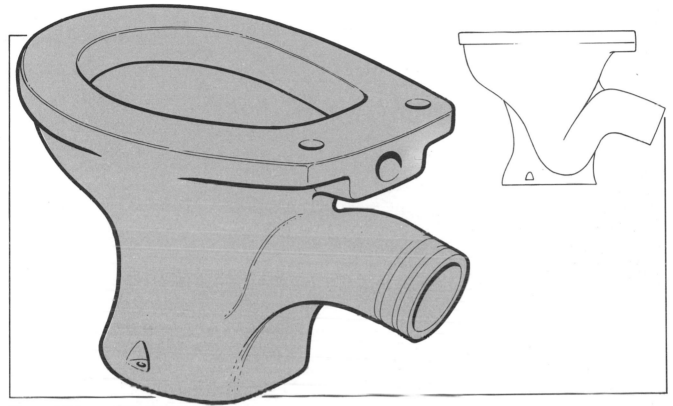

Fig. 171 A P trap as fitted to an upstairs
toilet

Opposite

Fig. 168 A corner-mounted seat bath
for the elderly or infirm

Fig. 169 A modern basin design

Fig. 170 Layout for basin and bidet

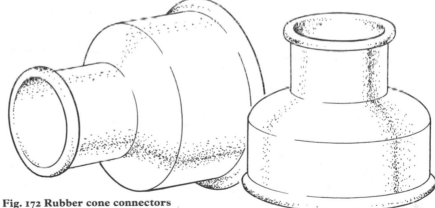

Fig. 172 Rubber cone connectors

CRACKED W.C. PANS

Never attempt to mend a cracked
toilet pan. Even if the crack is merely
superficial, any repair that may be
made will mean that the pan is no
longer a hygienic piece of equipment.
Renewing the pan is the only safe
remedy.

This job may appear daunting to the
inexperienced eye and with good cause,
especially if the toilet is on the ground
floor. The main problem here is how to
disconnect the cement joint between
trap and the drain socket without it
sustaining damage.

If the toilet is upstairs, however, it
will most probably be connected to a
soil pipe by means of a mastic joint.
This is less of a problem so I will deal
with this one first.

FLEXIBLE JOINT

The w.c. will discharge into the soil
pipe probably through a P trap upstairs, *fig. 171*. A flexible joint (mastic)
is used in these cases to allow for any
possible movement of the floorboards.
If the joint were solid cement, as may
be found on a ground floor w.c., the
pan could easily crack if there was any
noticeable movement.

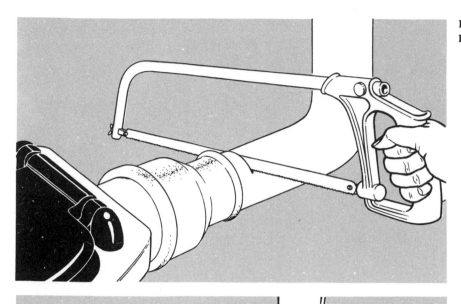

Fig. 173 Cutting a flush pipe with a hacksaw

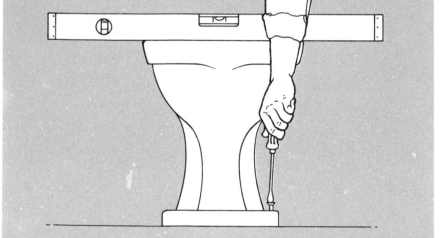

Fig. 174 Make sure the pan is level before screwing it down

Fig. 175 Multikwik connector

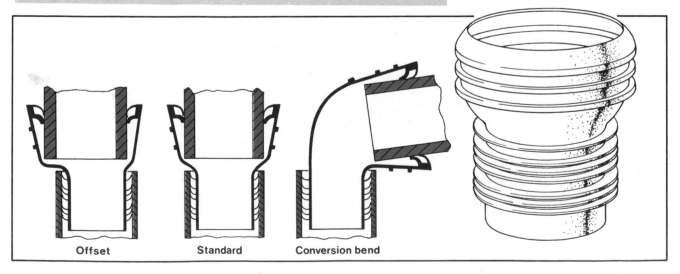

Offset   Standard   Conversion bend

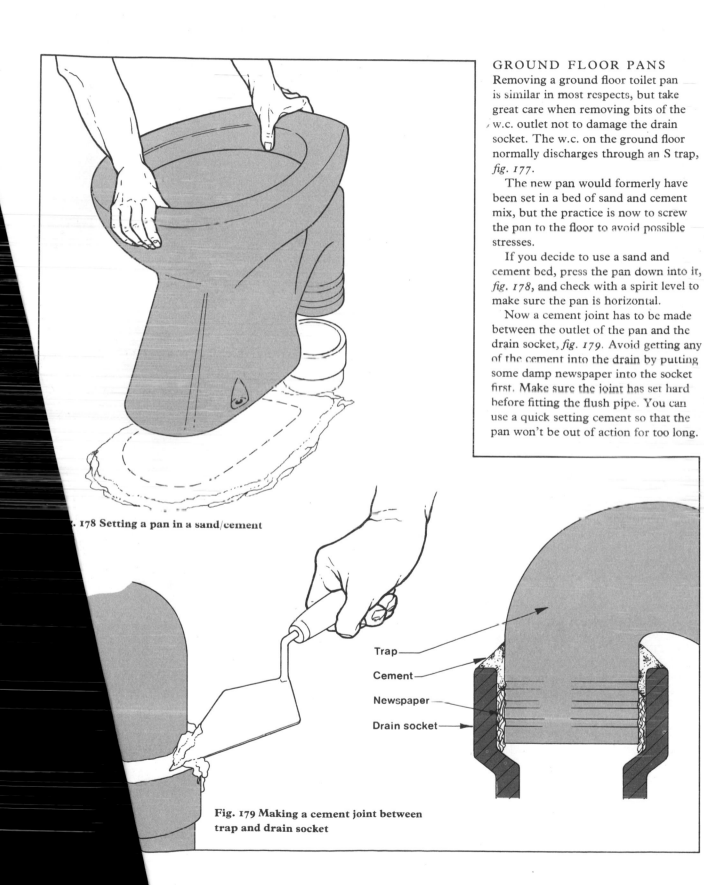

Fig. 178 Setting a pan in a sand/cement

Fig. 179 Making a cement joint between trap and drain socket

Trap

Cement

Newspaper

Drain socket

GROUND FLOOR PANS

Removing a ground floor toilet pan is similar in most respects, but take great care when removing bits of the w.c. outlet not to damage the drain socket. The w.c. on the ground floor normally discharges through an S trap, fig. 177.

The new pan would formerly have been set in a bed of sand and cement mix, but the practice is now to screw the pan to the floor to avoid possible stresses.

If you decide to use a sand and cement bed, press the pan down into it, fig. 178, and check with a spirit level to make sure the pan is horizontal.

Now a cement joint has to be made between the outlet of the pan and the drain socket, fig. 179. Avoid getting any of the cement into the drain by putting some damp newspaper into the socket first. Make sure the joint has set hard before fitting the flush pipe. You can use a quick setting cement so that the pan won't be out of action for too long.

To remove the old pan, first tie up the ball float arm in the cistern, flush and then disconnect and remove the flush pipe. If the joint holding this pipe is a rubber cone connector, *fig. 172*, simply roll this back and pull out the pipe.

If the joint is made with cement or putty, cut through the pipe with a hacksaw as close as you can get to the joint at the pan, *fig. 173*. If the pipe is lead or copper, I suggest it should be replaced with a plastic type.

Unscrew the nut under the cistern which connects the other end of the flush pipe.

You will now have to bale out the water lying in the trap. Then, with a sharp blow from a hammer, break the pan at the top of its bend. It will then be disconnected from the trap. You can then unscrew or lever the pan from the floor and remove it.

Take great care to prevent any rubble from now on falling into the soil pipe. Insert a thick wad of cloth so it fits tightly but can be easily removed. Carefully chip out all the remaining pieces of the old pan. Wipe the rim of the collar clean and remove the rag from the soil pipe.

GET THE PAN LEVEL

Now put the new pan in position and make certain it is absolutely level before screwing it to the floor, *fig. 174*. Use brass screws and don't overtighten them. You can fit washers or grommets under the screw heads to prevent damage to the pan.

Use a non-setting mastic such as Plumber's Mait to make the joint between the pan and the soil pipe. Bind Sylglas tape around the outlet, fill the area between the outlet and socket with mastic, then add another couple of turns of tape.

Alternatively, you can use a plastic connector called a Multikwik, *fig. 175*, between the w.c. outlet and a cast iron soil pipe or stoneware drain.

Now fit the flush pipe to the pan with a rubber connector, *fig. 176*, and connect its other end to the cistern.

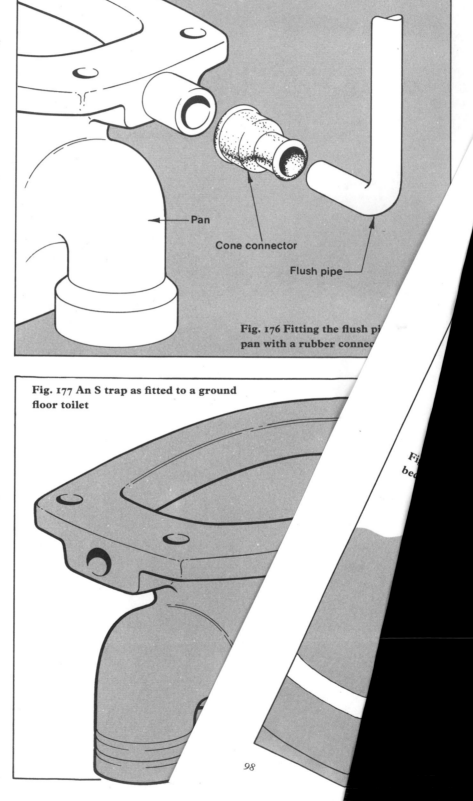

Pan

Cone connector

Flush pipe

Fig. 176 Fitting the flush pi pan with a rubber conne

Fig. 177 An S trap as fitted to a ground floor toilet

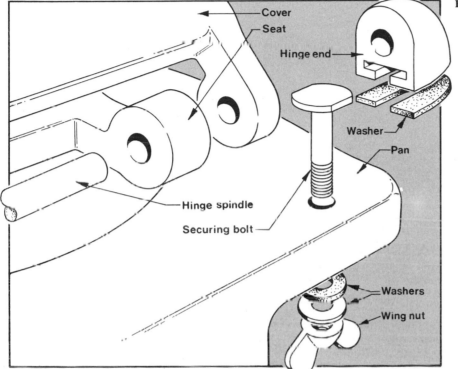

Fig. 180 One type of lavatory seat fixing

Cover
Seat
Hinge end
Washer
Pan
Hinge spindle
Securing bolt
Washers
Wing nut

TYPES OF W.C.

There are two main types of w.c. – siphonic and washdown.

In siphonic types the waste matter is removed by suction from a strong siphonic action. With washdown types, the waste is cleared by the volume and force of the flushed water.

Some siphonic pans are fitted with a single trap; others have double traps. The single types have a trappage shaped to retard the flushed water slightly until it completely refills the bowl of the trap. As the water moves towards the outlet, the siphonic action draws the contents of the pan into the soil pipe.

When a double trap siphonic pan is flushed, air is drawn from the pocket between the two seals by the flush water as it passes a special fitting. This causes atmospheric pressure to force the contents of the pan through both traps and into the soil pipe.

The traps are then resealed and the sides of the pan cleaned by water flowing from the rim.

An important point to watch when a new w.c. pan is contemplated is not to mix components made by different manufacturers. The flushing apparatus should be a suitable type for the installation concerned and should be bought with the pan as a complete unit or matched up, if that is possible.

NEW SEAT

When a lavatory seat or its lid becomes damaged, no attempt should be made to repair it; it should be renewed.

First measure up the size of the bolt holes at the back of the pan. Measure also the distance between the holes. The supplier of your new seat will need this information.

Seats are usually sold in units complete with lids, but you can sometimes buy them separately. Fitting instructions are usually supplied.

The units are simply bolted through the holes in the pan, *fig. 180*, and tightened with wing or ordinary nuts screwed down on to washers. Take great care not to overtighten the wing nuts. They should be only fingertight.

Fixing may vary slightly with make.

Central Heating

Although the subject of central heating is too broad and complex to be covered adequately in the space available here, many aspects of it fall within the sphere of plumbing.

Therefore, without going into great detail, I will try to give an outline of what a central heating system involves and pass on a few general hints regarding maintenance, installation and minor alterations.

First, how do central heating systems work? Their aim, of course, is to circulate warm air or hot water, or both, throughout the house from a central source of heat which is usually a boiler.

TYPES OF BOILER
There are basically three types of boiler – gas fired, oil fired or solid fuel. Apart from circulating heated water to radiators, some boilers also supply the hot water for domestic use. Most boilers used in domestic installations are free standing, but some open fires which burn solid fuel have a back boiler built in behind the fire.

Although boilers which are designed for central heating purposes are larger than those which supply the hot water for normal domestic use, they operate in much the same way. Some can be adjusted so that in summertime they heat only the domestic hot water and not the radiators which warm the rooms.

In gas fired boilers there are a number of burners that are lit by a pilot flame which should be on all the time.

A time switch is fitted to turn on the supply of gas automatically. This supply is controlled either by the boiler itself or by thermostats fitted in the rooms.

Where gas boilers are fitted in a room, some form of ventilation is needed to permit efficient combustion. Many gas boilers also require a flue. A chimney serves this purpose in many cases.

There is another boiler known as the balanced flue type. This is basically similar in principle to the conventional flued boiler, but some have fan assistance.

OIL FIRED BOILERS
There are various types of oil fired boilers. In one type – the *vaporising* boiler – the oil is fed into a pot. There it is heated, vaporised and burnt.

The oil in a *pressure jet boiler* is reduced to a fine spray and ignited by electricity.

Two other types – the *Dynaflame* and *Wallflame* – are fitted with electric motors which rotate jets that supply the oil. The oil is thrown against the burner ring where it is then reduced to small droplets.

Do not tinker with the controls of a boiler. This is a job for a trained engineer. If your central heating system is installed by a reputable company, they will make arrangements with you to have the system serviced at regular intervals.

I realise that many people have installed their own central heating and carry out their own maintenance. This is fine if you know what is involved. I have heard of many self-installed systems which work smoothly, but I have also heard of others which give constant trouble! So if you are thinking of installing your own system, do give it careful thought beforehand.

CONTROLS
Improvements are made constantly to the various controls used in a domestic central heating installation. Basically these controls can be divided into time and temperature types.

It is the time control which determines when the system is to be on or off. The most basic type is a time switch which has a dial marked off in 24 hourly sections. Around the dial are small levers which can be set manually to give either one or two on–off periods in the 24 hours as required.

A more advanced type is a programmer fitted with an electric clock. This type of control enables you to operate a programme for the hot water systems and the central heating.

TEMPERATURE CONTROLS
The simplest of these is a boiler thermostat which can usually be adjusted manually to vary the temperature of the water.

A room thermostat, *fig. 186*, however, gives better control of temperature. Some types are fitted with a clock and have a separate temperature setting for day and night.

In most cases a room thermostat is satisfactory for controlling the temperature of the water in the installation from a central point. But in larger installations, where the system may be divided into two or more zones, more than one thermostat will be needed. Under this arrangement, each thermostat will operate a separate circuit.

If a single room thermostat is to control the whole central heating system, it should be positioned so that it cannot give a false reading because of draughts or local heat. For example, don't stand a portable electric fire immediately beneath it!

OTHER CONTROLS
Most radiators are fitted with *lockshield* valves, *fig. 187*, which are operated by hand. These are for controlling the heat output of the radiators but are not the ideal means of controlling the actual temperature of the room. They are, however, invaluable for adjusting a system after it has been installed.

Where radiators are fitted with *thermostatic* valves, *fig. 187*, these will automatically control the room temperature at a predetermined level.

Mixing valves are also used to control water temperatures They mix some of the cooler water as it returns to the boiler with the hot water leaving. They also control the temperature of all the radiators connected to the system without any adjustment of the thermostat on the boiler being made.

WARM AIR SYSTEMS
Although the majority of central heating systems use radiators to circulate the heated water and thus heat the rooms, warm air systems use a

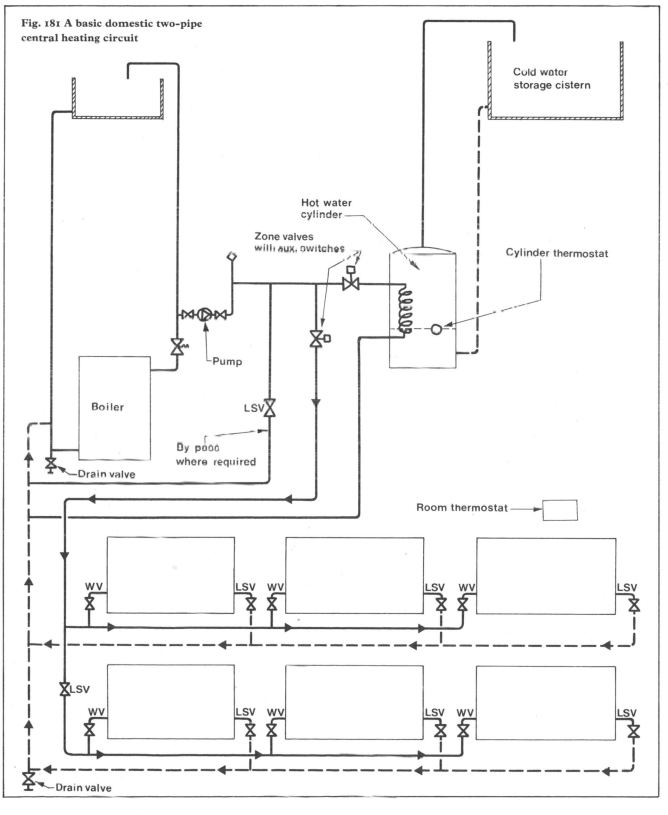

Fig. 181 A basic domestic two-pipe central heating circuit

Cold water storage cistern

Hot water cylinder

Zone valves with aux. switches

Cylinder thermostat

Pump

Boiler

LSV

By pass where required

Drain valve

Room thermostat

WV LSV WV LSV WV LSV

LSV

WV LSV WV LSV WV LSV

Drain valve

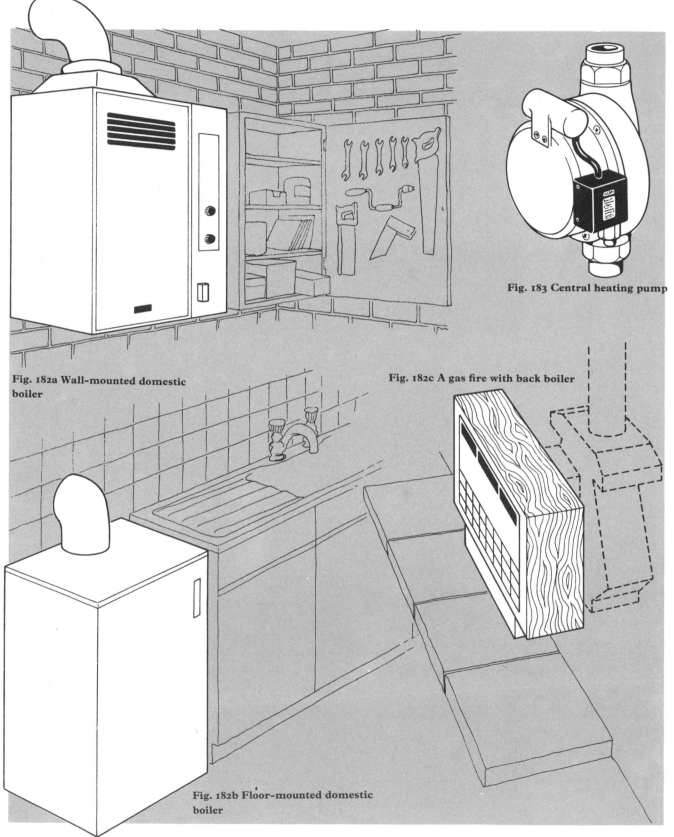

Fig. 182a Wall-mounted domestic
boiler

Fig. 182b Floor-mounted domestic
boiler

Fig. 182c A gas fire with back boiler

Fig. 183 Central heating pump

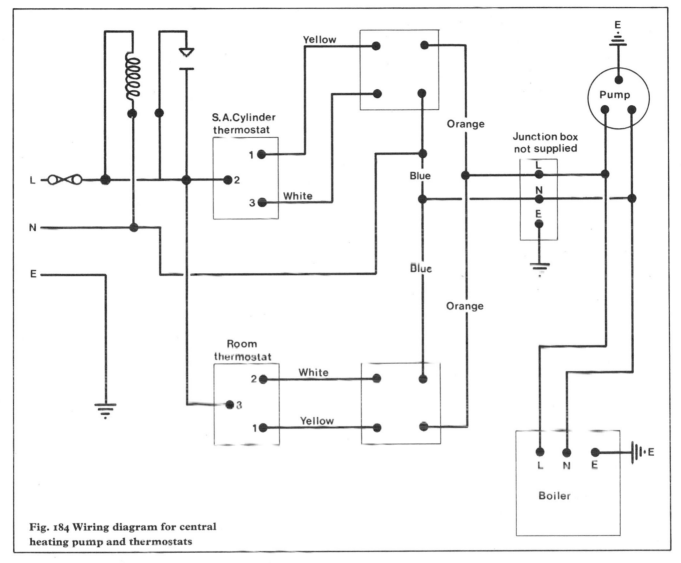

Fig. 184 Wiring diagram for central heating pump and thermostats

number of ducts through which air is circulated after being heated by electricity, gas, solid fuel or oil. In these systems the domestic hot water is usually supplied to the taps by an immersion heater or a separate boiler.

There are systems in which hot water rises from the boiler, passes through the radiators, and then, as it cools, returns by the force of gravity. These are known as gravity systems.

Usually, however, a pump is fitted near the boiler to circulate the water through the pipes and radiators. These are called forced circulation systems.

BOILER NOISES

Hissing – This can be a sign that the pipes, boiler and cylinder have become scaled up. See chapter on scale.

Another cause is overheating the water. Various things can be responsible for this: the flue may be blocked; a thermostat or the pump, *fig. 183*, may not be working properly; or there may not be enough water circulating in the system. The latter can be caused by the ball-valve in the feed and expansion tank becoming jammed.

If overheating occurs, shut off the gas or oil supply, or in the case of a solid fuel boiler, rake the burning fuel

into the ash pan and carry it outside. Then open the door of the boiler and let it cool down. Leave damper open.

If the pump can still be run while the boiler is switched off, let it pump water around to cool the system.

You can usually tell if a thermostat is doing its job properly. It should click when turned in either direction.

If the ball-valve in the expansion tank has jammed, it may need a new washer or simply dismantling and cleaning. See pages 17 and 18.

If the pump is switched on but no water is circulating, an air lock may be causing the trouble. Open the vent

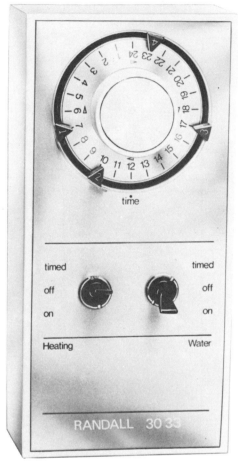

Fig. 185a Time switch controller for
pump and hot water

Fig. 185b Fully automatic programmed
control

Fig. 185c Time clock especially suitable
for single change-over circuits

Fig. 186 Room thermostat

valve of the pump, *fig. 183*, and bleed
the air off. Close the valve when the
water begins to flow again.

As a guide to temperature, that of
the water flowing from the boiler
should be about 7°C higher than the
water in the bottom pipe. You can test
the temperature by fixing a ther-
mometer first on the bottom pipe and
then on the upper one.

Adjustment of the flow regulator on
the pump should be left to a service
engineer.

ROOM TEMPERATURES
Well-designed central heating systems
keep room temperatures at a com-

fortable level by the use of room or radiator thermostats. Once these are adjusted to a satisfactory level, it should not be necessary to alter them.

No two people in a house have the same idea of a suitable temperature for a room, so a compromise figure will usually have to be agreed upon. The following temperatures are regarded as generally acceptable (all temperatures in Fahrenheit):

Bedroom 55°; living room 70°; bathroom 65°; hall, kitchen and toilet 60°.

DRAINING A SYSTEM

Before attempting to drain off a central heating system, make sure the fuel supply to the boiler is disconnected and that the ignition system is switched off. This should be done several hours beforehand to give the water time to cool down.

You will need a hose long enough to stretch from the draincock by the boiler to a drain outside.

First tie up the ball float arm of the feed and expansion tank. Fit the hose to the draincock and take the other end to the drain. Undo the draincock, *fig. 188*, and wait for the system to empty. Then close the draincock and release the ball float arm in the tank.

Let the system refill slowly. This will allow any air in it to escape through the vent.

AIR LOCKS IN RADIATORS

A common problem with central heating radiators is that sometimes they do not heat up as they should. Often the top of a radiator will be cool while the rest of it is warm. Air locks are a probable cause.

If this often happens, it is a good idea to fit an air eliminator, which works automatically, but the occasional air lock can usually be cleared quite simply.

A hollow key with a square-shaped end is usually required. This is inserted in the radiator, *fig. 189*, when the water is warm, and turned anti-clockwise to open the vent valve.

Some water may escape as the valve is opened, so hold a container underneath.

When the air stops coming out of the radiator and water begins to flow, tighten up the valve again. You may have to repeat the operation before a cure is effected.

The air eliminator has a valve containing a gland which is porous. This permits the air but not the water to escape.

Fig. 187 Lockshield valve and thermostatic valve

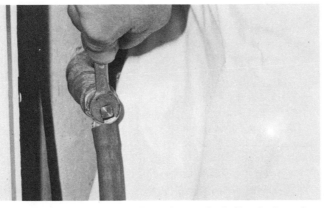

Fig. 188 To drain a central heating system, fit a hosepipe to the boiler draincock. Take the other end to a drain outside. Open the draincock with a spanner

Fig. 189 Clearing an air lock on a radiator

Fig. 190 Circulating pump with isolating valves fitted at each side for easy removal of pump

DRAIN THE SYSTEM

To fit an eliminator, the system will first have to be drained. Then undo the vent valve with the key as for an air lock, but this time remove the valve.

The eliminator is then screwed into the same hole. First, however, to ensure that the joint will be watertight, wrap some ptfe tape around the threads, *fig. 191.*

When you have screwed up the device fingertight, the system can be refilled.

If, after fitting the eliminator, you still find that a large amount of air needs to be released from the radiator, undo the adjusting screw of the device and hold a container beneath it to catch any water.

Take out the screw and gland.

Now press the valve with a screwdriver and bleed the system. Replace the screw and gland and turn the screw fingertight.

Bleeding the system is a job which should be done two or three times during the winter months whether air locks occur or not. When air gets into a radiator, it impairs its performance and there is always a possibility that internal corrosion may occur.

Bleed the ground floor radiators first.

LEAKS FROM JOINTS

Pipes and radiators in a central heating system expand and contract according to the temperature of the water. This can cause irritating leaks at the joints.

If a leak occurs around a compression joint, simply wipe the joint dry and tighten up its nut with a spanner.

When there is a leak at a capillary joint, however, enough water needs to be drained from the system to clear the water from the pipe.

Heat the joint with a blowtorch or blowlamp (mind the wall!) until you can release the pipe. Clean it up and apply flux to the end. Put the pipe back in the joint and solder around it until the joint is sealed.

Fig. 191 **For a watertight joint, wrap plastic tape around the thread of an air eliminator**

CARE OF THE BOILER

If you have a solid fuel boiler, clean out the ashes and top up with fuel regularly to keep the boiler burning steadily rather than fiercely.

Don't allow the boiler fire to die right down before adding more fuel to it. The boiler flue should be cleaned twice a year. Don't neglect the elbow of the flue near the base where deposits can settle. Give it a good clean out regularly

During very cold weather, the boiler should not be allowed to go out. If it *has* to be left cold for more than a few hours, drain the system completely and empty the boiler of fuel.

Do remember to refill the system before lighting the boiler!

Gas boilers are fitted with pilot lights and the majority contain a safety device which cuts off the supply of gas automatically if the pilot light should go out.

If at any time you should suspect a gas leak, turn off the gas at the main cock which is fitted near or on the meter. Don't try to find the leak by testing with a naked flame! Telephone your local gas board at once.

OIL STORAGE TANKS

If you have an oil-fired boiler, the oil storage tank should be maintained regularly. There are small things you can do to ensure that it works smoothly.

It is important to keep the filter clean. Turn off the stopcock, remove the filter bowl and clean the element with petrol. Reassemble it when dry.

Keep the cap of the filter pipe oiled so it can be removed easily.

The vent pipe must be kept clear of obstructions. You can fit wire mesh over it to keep out leaves or other debris. If the vent pipe does become obstructed, clear it with a length of stiff wire.

Sludge can form in the storage tank. This should be drained off about once a year by opening up the draincock to let the sludge run out. Stop draining when pure oil begins to flow.

Exposure to the elements may cause the sides of the tank to rust. Any rust should be removed by wire brushing and abrasive paper. Then coat the tank with a rust inhibitor followed by a coat of bituminous paint.

Do this regularly and pay particular attention to the welded seams.

FLUES

Most gas, solid fuel and oil-fired boilers need a flue. For the best results, this should be as straight as possible. Sharp bends should be avoided as they hinder the flow of gases and allow debris and soot to collect.

If oil or solid fuel appliances are used, the height of the flue becomes especially important to ensure adequate draught. Generally speaking, a tall chimney will provide more draught than a short one.

Height of the chimney is, of course, less important for gas appliances. Here the flue acts as a channel or duct to carry the combustion products out into the open air.

An ordinary brick chimney by itself is not always suitable for this purpose. This is because very little heat escapes up chimneys when modern gas appliances are used.

Consequently, as the gases pass through a relatively cold chimney, they cool down to such an extent that condensation can form. This in turn can lead to damp walls.

In these situations, the flue is usually lined and the best method is to have a flexible stainless steel liner inserted in the chimney, *fig. 192*. Your heating

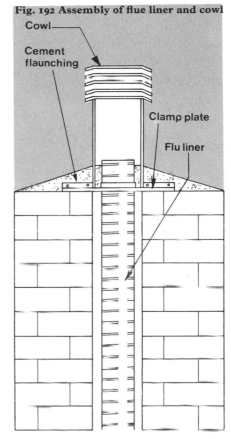

Fig. 192 **Assembly of flue liner and cowl**

Cowl

Cement flaunching

Clamp plate

Flu liner

supplier should be able to arrange this for you if it is necessary.

A terminal is normally fitted to the top of the chimney to stop birds nesting and to keep out the rain. Examples of gas and oil terminals can be seen at your suppliers.

RADIATOR POSITIONS

While it is important that any central heating system should be carefully designed to give maximum efficiency, the positioning of radiators does not always get the attention it deserves.

To lessen the affects of cold draughts entering, radiators should, if at all possible, be positioned underneath windows – preferably running along their full length. If this is not convenient, place the radiator on a wall adjacent to the window.

Don't position a radiator opposite windows. This could make the draughts even more noticeable.

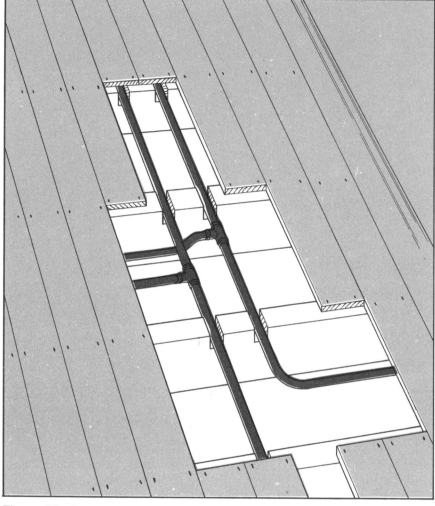

Fig. 193 Pipe layout recessed into joists

The best type to use in this situation is a fan convector radiator. This will distribute warm air across the window area and help to counter the effect of the incoming draughts.

If radiators have to be fixed on outside walls, some of their heat may be lost *through* the walls. This can be prevented to a degree by insulating the internal wall surface, perhaps by hanging sheet foam polystyrene under a heavy vinyl wallcovering.

Alternatives are to attach insulation boards and to decorate over them, or use aluminium foil. This treatment is more suitable for solid walls. For cavity walls, of course, specialist applied cavity infill is the best solution.

OBSTACLES

A point to remember about a radiator is that a great deal of the heat it generates is the convected type. A smaller proportion of radiant heat is produced.

Single-panel radiators produce more radiant heat than two- or three-panel radiators. Often, however, the full benefit of this heat is not obtained because large items of furniture are placed in its path. If this can be avoided, so much the better.

D-I-Y KITS

For the do-it-yourself enthusiast who wants to install his own central heating, various firms produce kits designed especially for this purpose. Many of these companies also offer design services to cover all aspects of the work.

I need hardly add that to do this job and make a success of it, you need preferably at least some plumbing experience. Equally essential is a sound understanding of what is involved.

It is worth paying the fee that most firms charge for their design services. There are many things to be considered such as correct balancing of the system, the outputs and inputs of various appliances and the complicated business of U values and heat losses.

These things are all taken care of by reputable firms who provide such design services. All you have to worry about is the physical job of installation! This, of course, is an involved subject beyond the scope of this book, but here are some general hints which I hope will prove useful.

If you deal with a reputable firm you should get full instructions on how to assemble all the components.

ORDER OF WORK

Advance knowledge, however, is always useful. Some firms advertise very comprehensive catalogues which contain many helpful illustrations or photos of the various bits and pieces. It is well worthwhile following the advertisements and sending for one or two of these catalogues.

If you order a kit to assemble yourself, check it over carefully when it arrives and familiarise yourself with all the parts. As I said, the kit should contain full instructions for assembly, but here is a general guide to the order in which the various jobs are usually done, though much will, of course, depend on the system.

1. Fit the radiator valve unions, plugs and air cocks.

2. Fix the radiator brackets to the wall.

3. Fix the radiators on the brackets and then add the radiator valves.

4. Assemble the boiler.

5. Fix the fittings to the hot water

cylinder and position it.

6. Attach the fittings (outlet, overflow and ball-valve) to the feed and expansion tank.

7. Mark out the runs for the pipes. Drill holes in the walls for the pipes as necessary.

8. Run in the pipes.

9. Fix the pump.

10. Connect the boiler to the chimney.

11. Connect up the oil or gas supply to the appliance. The gas board should make a check at this final stage.

FITTING THE BOILER

The boiler must stand on a perfectly flat and level base and the centre of the flue outlet must line up with the centre of the flue itself.

Make sure that all joints, especially those on the boiler, are absolutely watertight.

When fixing the fittings on to the hot water cylinder, avoid overtightening them and don't confuse the connections. Note that the male threads are the primary connections and are connected to the boiler. The female threads are those for the domestic hot water connections.

Use ptfe tape on the male threads to ensure that the connections will be watertight.

USE WASHERS

When you come to drill holes in the feed and expansion tank for the various fittings (ball-valve, overflow and cold feed), don't use jointing compound if the tank is a plastic type. Here you should use polythene washers.

To make holes in walls for the pipes a masonry bit used in a power drill and driven at a low speed is better than using a hammer and chisel – and less noisy.

There are two important points to note when running in the pipes. Keep the primary pipes rising steadily throughout their journey and ensure that vent pipes rise continuously from the point where they are connected.

Bending pipes is dealt with else-

Fig. 194 Flue terminal

where. Avoid making too many bends in one length of pipe. For one thing, the pipe will look unsightly; for another, it will be difficult to manoeuvre into its final position. Although it works out more expensive, it is better to use compression fittings to achieve neat changes in pipe direction.

FILLING THE SYSTEM

When the proud moment comes to fill the system, do this slowly so that you can detect and remedy leaks before any damage is done.

If you have had to lift floorboards in order to lay pipes, it is advisable not to replace the boards until the system has been filled and checked. Then I suggest you replace the boards with screws instead of nails so that future access will be easier.

The gate valve on the domestic cold supply pipe should be opened a couple of turns to enable the cylinder to fill gradually.

As soon as water begins to flow from the hot taps, turn them off. The cold water cistern can then refill.

Now turn on the cold supply pipe to the feed and expansion tank and make sure all the radiator valves are open.

Keep a weather eye open for leaks in the pipework while the system is being filled.

BLEEDING

When the system is full you will need to bleed the radiators, boiler and pump which may be filled with air.

When you are sure the system is free of leaks, drain it and flush through a couple of times before finally filling it.

The next job will be to fire the boiler, but first make certain that all the controls and valves on the radiators are open.

Set all thermostats as required and make sure that the pump is working. Here you should pay particular attention to the instructions supplied.

Now the system can be balanced. Before this is done, however, the lockshield valves on the radiators should be partly closed. Balance the radiators when they are hot.

While doing this, the room thermostat should be set at maximum and the pump adjusted to give the flow required.

CORROSION

There are a number of factors which can cause corrosion in a central heating system. One example is the use of different metals for the components and pipework.

Gases are produced in most systems and can cause air locks. They also create deposits which collect in the pipes and radiators. These can restrict the circulation of water.

These deposits are in themselves a form of rust and are magnetic. Consequently they are attracted by the circulating pump because of a magnetic field which the electric motors of the pump produce. If this is allowed to get out of hand, pump failure is possible.

This can be avoided by introducing a chemical inhibitor into the system, and this is a point worth discussing with your supplier at the outset.

COMMON PROBLEMS

Corrosion in one form or another is a common cause of trouble in central heating systems. In a *direct* hot water system, corrosion can, of course, occur very soon after installation. In an *indirect* system, the process is considerably slower, but it can happen.

Air is an essential element in the process of corrosion. As a direct system is constantly filled with fresh aerated water, this can corrode rapidly.

An indirect system has a closed heating circuit. Once it is filled, the same water is used over and over again (unless the system is drained, of course). The system's water, therefore, becomes stale once the air in it dissolves.

It would seem, then, that with an indirect system corrosion cannot take place. Unfortunately, though, it is possible because air can enter the system in other ways – at the feed and expansion tank and through tiny leaks at pipe joints. These are too small to let any water out but are big enough to let the air get in.

SYMPTOMS

If the system is not protected by a chemical corrosion inhibitor, the corrosion will sooner or later cause various problems. These include: radiators running cold and requiring frequent venting, small leaks in the radiators and pump failure.

After a time it may be found that the radiators are not warm all over. If this is discovered after the system is vented, it may be that the radiators are clogged with sludge, which is also due to corrosion. This sludge will eventually find its way around the system to the pump and cause it to fail.

The chemical inhibitors can be introduced into the system via the feed and expansion tank to prevent this happening. Two well-known inhibitors are Fernox and Radicon. Your supplier should be able to suggest one that is suitable.

HUMIDIFIERS

After their central heating has been installed many people find that the atmosphere in the house becomes too dry for comfort. They complain of sore, dry throats, headaches and possibly sinus troubles.

Sometimes even the furniture is affected; the adhesive at the timber joints dries out. Pianos have even been known to become unplayable. The better the central heating system, the more prevalent this sort of trouble is likely to be.

The answer is to fit a humidifier which will ensure that moisture in the air will be kept at the proper level.

Atmospheric moisture is determined by the relative humidity. For example, in a centrally-heated room which has no humidifier and a temperature of $70°F$, the relative humidity will be around 20%. The recommended humidity is nearer 55%.

Some form of humidification is needed to provide the necessary water vapour and the figure of 55%.

Some people stand saucers of water in the room to moisten the air, but this is not the solution to the dry air problem. A proper humidifier is needed which has the appropriate output for the situation.

CALCULATIONS

To get the exact amount of humidification, various calculations need to be made. These involve the size and layout of the room, the air temperature and that of surfaces, plus the number of air changes made per hour in the room. To get it right you need expert advice.

There are two sorts of humidification, background and positive. The former prevents dry air absorbing moisture from the surroundings. It cannot, however, restore the humidity to the recommended level of 55%.

Positive humidification, however, *can* add the correct amount of moisture to the dry air.

To achieve this, you need an electric type of humidifier which is capable of producing a minimum of half-a-pint of water vapour every hour. It should be fitted with an automatic humidistat (this is rather like a thermostat) and a means of adjusting the output of vapour.

There are several types on the market and there is a specialist company which can give you advice on the subject – The Humidifier Advisory Service, Felvic House, 21 Napier Road, Bromley, Kent BR2 9JA.

Improvement Grants/Plumbing Terms

If your property lacks a modern plumbing system, you may be able to get a home improvement grant towards the cost of installing up-to-date amenities. Much will, of course, depend on your council, on the property itself and on the local bylaws.

At the time of writing, three types of grant are available:

1. Standard grants for the provision of standard amenities (a bath or shower; washbasins; sink; a hot and cold water supply to a fixed bath or shower, washbasin or sink; a w.c.).

2. Improvement grants to improve the standard of existing properties, or to convert them into flats. These grants do not apply to modern houses or properties which are well equipped and in good repair.

3. Special grants at the discretion of local councils for standard amenities to be shared in houses in multiple occupation, if there is no immediate likelihood of converting the property into permanent separate dwellings.

To find out more about home improvement grants, apply to your local council. Whatever you do, don't start work on a project in anticipation of getting a grant. Any scheme you may have in mind *must* be approved by the council first. They will not entertain an application for a grant if the work has already begun.

There are hundreds of terms used in the plumbing industry and many variations of some of them. For example, the term stopcock is used in many cases when in others it might be called a stop-valve or stoptap.

So far as the householder is concerned, he need not concern himself about the variations of these terms, so long as he can make his requirements understood when making purchases.

Here are a few terms which may be encountered, with a brief definition of each.

Air lock Pocket of air trapped in pipe. Tap splutters and may stop running.

Backnut Nut on stem of fitting (a tap, for example) for fixing to basin or other fixture.

Ball valve Valve which controls the flow of water into a cistern or tank.

Bib tap Tap with horizontal inlet. Sometimes referred to as a bib cock.

Bird's mouth End of an overflow pipe shaped like a bird's beak.

B.Th.U. British Thermal Unit. Amount of heat needed to raise 1 lb of water through one degree Fahrenheit.

Crow's foot Basin wrench or spanner. Can be used vertically or horizontally.

Downpipe Rainwater pipe.

Drift Tool for opening out the end of a copper tube. Also called swaging tool and flaring tool.

Elbow Fitting which enables a change to be made in pipe direction.

Flashing Strip of lead, copper or zinc which covers a joint between a roof covering and another part of the building.

Interceptor Trap on drain to stop sewer gases entering.

Jumper Combined disc and spindle to which tap or stop valve washer is fixed.

Mandrel Cylindrical length of hardwood to shape lead pipes.

Pillar tap Tap with a vertical inlet.

Rising main Pipe which carries the water from the main into a house and up to the cold water storage cistern.

Seal Water in a trap under a sink, basin, bath or w.c. which prevents the passage of foul air.

Spigot Plain end of a pipe or gutter which is fitted into the enlarged end (faucet) of the following section.

Stopcock Device fixed in a pipe to regulate the flow of water.

Stop-valve A valve fitted in a pipe to control the flow of water.

Note: both stopcock and stop-valve serve the same purpose. They are described variously in different parts of the country.

Swan neck A pipe bend (offset). The pipe continues on a parallel course.

Trap Fitting or part of an appliance which stops the passage of foul air.

Water hammer Knocking noises in pipes caused by high water pressure. Indicates a fault in the system.

Index

Additives, hot water system 74
Air eliminator, radiator 106
Air extractors 79
Air locks 24
  preventing 25
  radiator, curing 105
Anti-siphonage vent pipes 64
Asbestos cement cisterns 9

Ball-float arm, adjustable 17
Ball-float troubles and repair 17
Ball-valves 16, 19
  diaphragm type 18–20
  equilibrium 20
  faults 17
  flushing cistern 26
  fitting, 15
  heater tank, 23
  noisy 19
  pressures 18
  rewashering 18
  storage cistern 9
Bath:
  boxing in 90
  new, fixing 89
  old, removing 88
Bathroom condensation 79
Bib tap 32
Bidets 91
Bleeding heating system 105, 106, 109
Blockage, clearing a waste-pipe 57–60
Boiler 56, 100, 107
  fitting 109
  noises 103
Boiling point safety device 22
Bottle trap 62
  rewashering 64
Burst pipes 54
  preventing 50–56
  repairing 55

Capillary joints 45–47
  remaking 106
Calorifier, 22
Central heating 100–112
  boiler, fitting 109
  controls 100, 105
  d-i-y kits 108
  draining 105
  filling the system 109
  flue 107
  pump 102, 103
  warm air 100
Chimney, lining 107
Cistern(s) 7, 26
  arrangements 7
  capacities 11, 13
  flushing problems 26
  lagging 50, 53, 54
  low level 27
  polythene, installing 15
  repairs 76
  water level, adjusting 17, 29
Cold water system 7–20
  supply points, 9, 10
Compression joint:
  manipulative 43–45
  non-manipulative 41–43, 47, 49
Condensation 53, 78
Conversion table 49
Copper pipes and joints, 41–47
Corrosion 75, 76, 110
  inhibitors 110

Crow's foot spanner 39, 111
Croydon ball-valve 16
  rewashering 18
Cylinder, hot water:
  collapse, 56
  direct, 21
  indirect, 22

Descaling hot water system 74
Direct cold water system 9
Direct hot water system 21
  scale in 74
Double glazing and condensation 79
Drain rods 59
Drainage systems:
  one-pipe 62, 64
  two-pipe 59, 60
  ventilating 61
Draincock 7, 8, 21, 22

Electrical heating tape for frost protection 54
Endfeed capillary joint 47
Equilibrium ball-valve 20
Expansion tank 23

Fan heater for frost protection 53
Filling heating system 109
'Fix-a-Tap' set 40
Flap valve, W.C., renewing 28
Flues 107
Flushing cistern 26
  low level type 27
  problems 26
Fractured pipe in ground 59
Frost precautions 50–56
'Full Stop' tap washer and seating 35
Full-way gate valves 37

Gas-fired boilers 100, 107
Gate valves 9, 11, 37
Gland packing, water tap 36
Gullies, cleaning 59
Guttering 63
  cast-iron, replacing 67, 68
  plastic, fitting 69

Header tank 23
Heaters for frost protection 52
Heating tape, electrical 54
Heating, warm air 100
Hopper, waste-pipe 60
Hot water systems:
  air locks in 24
  direct 21, 74
  indirect 21
  refilling 25
Hot water tanks, repairing 76
Hot water taps, rewashering 35
Humidifiers 110

Improvement grants 111
Indirect cold water system 9
Indirect hot water system 21–23, 75
Insulation, house 79
'Isopon' 55, 65

Joints, pipework:
  capillary 45
  compression, manipulative 43–45
  compression, non-manipulative 41–43, 47, 49
  leaking, dealing with 106

Jumper, tap:
  pegged 35
  removing 33, 34

Lagging plumbing work 50–54

Manhole 60
  vent pipe 61
'Mapel Tank-Saver' 76
'Markfram' W.C. control valve 20
'Metalife' solid tank saver 76
Metrication 49
'Micromet' additive for hot water systems 74
Mixer valves for showers 81
'Multikwik' W.C. connector 97

'Neverstop' W.C. control valve 20
Noisy ball-valves 19

Oil-fired boilers 100
Oil storage tank 107
Overflow pipe 15
  protection against frost 51, 52, 54
Overflow, stopping an 17
Overheating, causes of 103

Painting:
  cast-iron gutters 67
  cold water cistern 75
  plastic gutters 69
Pillar tap 32
Pipe bending:
  copper 71
  polythene 73
Pipe joints:
  capillary 45–47, 106
  compression 41–45, 47, 49
  Imperial and metric conversions for 49
  rubber cone connector 97
Pipes 7
  copper 41
  leaky, mending 55, 106
  polythene 49
  rigid PVC 49
  stainless steel 47
Plastic padding 55, 65
'Plumber's Mait' non-setting mastic 97
Plumbing terms 111
Polythene cistern 9
  installing 15
Polythene pipes 49
Portsmouth ball-valve 16
  rewashering 18
Putty, metal casement 67

Radiators 107
  air locks, removing 105, 106
  feed for 23
Rainwater fittings, plastic 69
Refilling hot water system 25
Rigid PVC tube 49
Rising main 7, 8
Rust 75
  in gutters 65
  removal, 67, 76, 110

Sacrificial magnesium anodes 76
Scale in hot water system 74
Shower, 80
  circuit 83
  heads 81

Showers, continued
  installing 83, 84
  mixing valves 81
  types of 82
  water heater 81
  water pressure 80
Silencing cisterns 19
Soil pipe cage 68
Soldered capillary joints 45–48
Solid-fuel boiler 107
Stainless steel tube 47
Stopcock 7, 8, 37
Stop-valve 7, 8
  and draincock combined 8
  fitting 43, 71
  gateway 9, 11
  rewashering 37
  W.C. inlet pipe, for 20
String line for gutters 69
Supatap 34
  rewashering 35

Taps 36, 37
  fitting 38–40
  loose, curing 36, 37
  outside, installing 70–73
  repacking gland 36
  rewashering 30–35
  worn seating, dealing with 35
Temperature controls 100, 105
Thermostatic valves for showers 81
Toilet blockage 58
Toilet cistern 26
  flushing problems 26
  low level 27
Toilets, outside 54
Traps:
  waste 62, 64
  waste, rewashering 64
  W.C. 95

Underground pipe blockage 59, 60

Vent pipes 109
Ventilating house drains 61
Vermiculite granules 53

Wall surface, insulating 108
Warm air central heating 100
Washbasins 92, 93
Washers:
  for plastic tank fittings 109
  tap 30–35
  trap 64
  W.C. ball-valve 16–18
Waste fitting, sealing a 58
Waste pipes 60
  blockage, clearing 57
Waste traps 62
  rewashering 64
Water closet 99
  pans, downstairs, replacing 98
  pans, upstairs, replacing 95
  seat, renewing 99
  traps 95
Water hammer 19
Water heater for shower 81
Water pressure for shower 80
Water level in cistern, adjusting 17, 29

Poems of the Seasons

PICTURES BY
GORDON BENINGFIELD

This edition published by Selectabook Ltd,
Folly Road, Roundway, Devizes, Wiltshire SN10 2HT
produced by TAJ Books,
27 Ferndown Gardens, Cobham, Surrey, KT11 2BH

www.tajbooks.com

Selected and designed by Jill Hollis and Ian Cameron
Cameron Books, PO Box 1, Moffat, Dumfriesshire DG10 9SU

www.cameronbooks.co.uk

First published 1992
Copyright © the Estate of Gordon Beningfield (pictures) 1978, 1980, 1983, 1985, 1987, 1988, 1989, 1990, 1992
and Cameron Books (design and this selection of poetry)

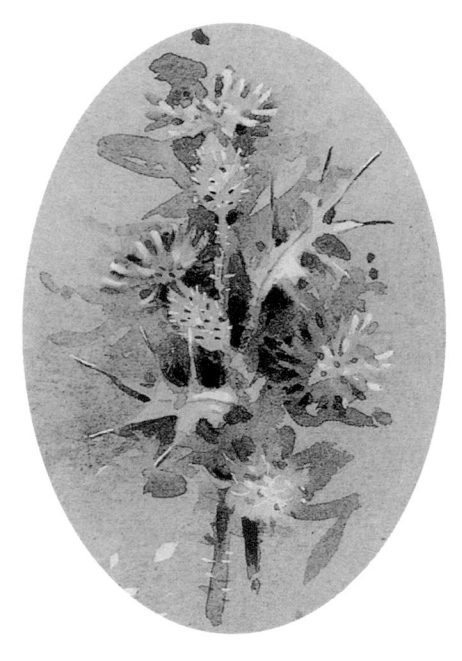

ISBN-1-84406-006-3

Reprinted 2003

Thanks are due for their permission to reproduce copyright material in this volume: to Mrs
Nicolete Gray and the Society of Authors on behalf of the Laurence Binyon Estate for
'Early June', 'Harebell and Pansy', and extracts from 'Strange Fruit' and 'Thunder on the
Downs' by Laurence Binyon; to Peters Fraser & Dunlop Ltd for extracts from 'The Gift: for
CMP' and 'Wilderness, and to Carcanet Press Ltd for 'What is Winter?' by Edmund
Blunden from Selected Poems ed. Robyn Marsack; to Rosemary Seymour for 'Sudden Spring'
by Gerald Bullett; to the Estate of Richard Church for 'A Procession' by Richard Church; to
the Literary Executor of Leonard Clark for 'Christmas' by Leonard Clark; to the Literary
Trustees of Walter de la Mare and The Society of Authors as their representative for
'Martins: September' by Walter de la Mare; to the Estate of Robert Frost for 'Reluctance'
from The Poetry of Robert Frost ed. E.C. Lathem, published by Jonathan Cape; to the
Enitharmon Press for 'St Luke's Summer' by Phoebe Hesketh; to Faber and Faber Ltd for
'April Birthday' by Ted Hughes; to Peters Fraser & Dunlop Ltd for 'April Rise' and 'Cock
Pheasant' by Laurie Lee; to Curtis Brown, London, for 'Fritillaries', an extract from The
Land copyright 1947 Vita Sackville-West; to Alison Young (© The Andrew Young Estate) for
'A Windy Day' by Andrew Young. Cameron Books has made every effort to obtain
permission to reproduce material in copyright and apologises to any copyright holder who
has proved impossible to contact.

SELECT
EDITIONS

Pictures

Kingcups	endpapers	Bluebell word	47	Harvest scene	87
Dairy Shorthorn	frontispiece	Forget-me-nots	48	Rhode Island Reds	89
Blackberries	title page	Bulrushes	49	Rick yard	90
Thistle	6	Watermeadows in the Frome valley	51	Loading sheaves	93
Heathland	10	Dorset Horn sheep	52	Apples	95
Snowdrops	13	Dandelions	55	Swallows	97
The River Hook	14	Honeysuckle	57	Pump mill and gulls	98
The lambing yard	16	Flowers in lawn	59	Basket of apples	100
Hampshire Down sheep and lambs	19	Red clover	60	Cock pheasant	103
Primroses	21	Bumblebee	61	Collapsed haystack	104
Farm gate	115	Toller-Down	63	Old man's beard and rose hips	106
Heavy horses	24	Foxglove	64	Haws	108
Landscape with rooks	27	Little Gaddesden church and stile	65	Hazelnuts	109
Celandines	29	Sussex Downs	66	Scarecrow	110
Pussy willow	30	Wild roses	69	Black bryony	111
The Third Kissing Gate	33	Sturminster Newton Mill	70	Autumn leaves	113
Whitepark and calf	35	Skylark	73	Little Gaddesden and beech trees	114
Violets	37	Marble White and Small Skipper	74	Shepherd's hut with lambing fold	116
Nest	38	Harebells	75	Dorset barn	119
Hazel catkins	39	Dorset sky	76	Holly and mistletoe	121
Thrush on hawthorn	41	Small Tortoiseshells	79	Cart track with partridges	122
Lady's smock	42	Mill on the Bure	81	Tring reservoir	124
Fritillary	43	Dogbury Hill	82		
Apple orchard	44	Poppies	85		

Poems

John Armstrong *The Seasons* — 11

Matthew Arnold extract from *Thyrsis* — 65

Evelyn D. Bangay *On Ploughing* — 26

George Barlow *Bluebells* — 46

Laurence Binyon *Early June* — 53
    Harebell and Pansy — 74
    extract from *Strange Fruit* — 101
    extract from *Thunder on the Downs* — 83

William Blake extract from *To Autumn* — 101

Edmund Blunden extract from *The Gift: for CMP* — 49
    What is Winter? — 117
    extract from *Wilderness* — 106

Lilian Bowes Lyon *Evening* — 99

Nicholas Breton extract from *The Passionate Shepherd* — 31

Robert Bridges extract from *A Defence of English Spring* — 42
    extracts from *The Months* — 61, 94
    extract from *North Wind in October* — 109
    extract from *The Storm is Over* — 113

William Broome *The Sun* — 77

Robert Browning extract from *Misconceptions* — 40

Robert Buchanan *Snowdrop* — 13

Gerald Bullett *Sudden Spring* — 32

Geoffrey Chaucer extract from *The Legend of Good Women* — 59

Richard Church *A Procession* — 61

John Clare extract from *The Eternity of Nature* — 59
    The Flitting — 62
    extract from *On May Morning* — 48

Leonard Clark *Christmas* — 120

Hartley Coleridge *November* — 112
    The Snowdrop — 12

Samuel Taylor Coleridge *The Flowering May-Thorn Tree* — 40
    extract from *Frost at Midnight* — 125
    extract from *The Keepsake* — 48

William Cowper extract from *The Task* — 123

Adelaide Crapsey *November Night* — 117

George Darley extract from *Song of the Bluebells* — 46

Walter de la Mare *Martins: September* — 96

Emily Dickinson *The Bee* — 60

William Dowsing *There's Many a Flower* — 36

Michael Drayton extract from *Ballad of Dowsabel* — 57

John Drinkwater *The Miracle* — 23

John Dyer *The New-Drop'd Lamb* — 18

Richard Edwards *The Mood of May* — 50

John Fletcher *The Falling Night* — 17

John Freeman *Evening Orchard* — 45
    Last Hours — 115
    November Skies — 117
    The Unloosening — 15

Robert Frost *Reluctance* — 118

John Gay *The Careful Insect* — 60

Thomas Hardy *The Last Week in October* — 112

Robert Herrick *To Violets* — 37

Phoebe Hesketh *St Luke's Summer* — 108

Gerard Manley Hopkins *Sky* — 80

Henry Howard *Description of Spring* — 40

Mary Howitt *The Voice of Spring* 39

Ted Hughes *April Birthday* 30

James Hurnard extract from *The Setting Sun* 121

Helen Hunt Jackson *Poppies in the Wheat* 84

John Keats *A Day Born of the Gentle South* 94
    extract from *'I Stood Tip-toe Upon A Little Hill'* 40
    Summer Night 68

John Keble *Spring Flowers* 20

Frank Kendon *Bright Autumn* 97

Sidney Keyes *The Pheasant* 102

Mary Leapor *Autumnal Threads* 105
    My Father's Fields 92
    A Summer's Day 56

Laurie Lee *April Rise* 32
    Cock Pheasant 102

James Russell Lowell *To the Dandelion* 54
    extract from *The Vision of Sir Launfal* 71

George Macdonald *To—* 68
    The Sun 77

Lucia C. Markham *Bluebells* 46

George Meredith *The Harebell* 75
    extract from *Seed-Time* 107

William Morris *Autumn* 111

David Morton *These Fields at Evening* 99

Edith Nesbit *Winter Violets* 36

Caroline Elizabeth Norton *Harbingers of Spring* 13

Matthew Prior *Nesting Birds* 38

Mary Eleanor Roberts *The Heaven-Soaring Lark* 73

Edwin Arlington Robinson *The Sheaves* 86

Christina Rossetti *On a Bed of Forget-me-nots* 48
    A Green Cornfield 72
    extract from *Seasons* 92
    Summer 67
    extracts from *A Year's Windfalls* 39, 45, 95

Margaret Sackville *Vanished Summers* 88

Victoria Sackville-West *Fritillaries* 43

William Shakespeare *Blow, Blow, Thou Winter Wind*
    from *As You Like It* 123
    Honeysuckle from *Much Ado About Nothing* 57
    A Violet from *Hamlet* 37

Percy Bysshe Shelley *The Primrose* 20

Charlotte Smith *At the Close of Spring* 53

Margaret Stanley-Wrench *Ploughing up the Pasture* 25

Alfred, Lord Tennyson *By a Brook* 34
    extract from *Gareth and Lynette* 57
    extract from *The Lotos-Eaters* 94
    extract from *The May Queen* 42

Edward Thomas *The Brook* 78

Francis Thompson extract from *The Poppy* 84

James Thomson extracts from *The Seasons* 15, 60, 77, 91, 92

Jones Very *The Latter Rain* 105

Edmund Waller *The Lark* 73

John Byrne Leicester Warren extract from *Auguries of May* 49

John William Watson *April* 31

Rosamund Marriott Watson *Traveller's Joy* 106

Augusta Webster *Holly* 120
    The Violet and the Rose 37

Henry Kirke White *To an Early Primrose* 20

Nathaniel Parker Willis extract from *The Month of June* 68

William Wordsworth extracts from *The Prelude* 64, 75
    September 1815 86
    To the Daisy 58
    To the Small Celandine 29
    To a Snowdrop 12

Andrew Young *A Windy Day* 110

The Seasons

from *The Art of Preserving Health*

Observe the circling year, how unperceiv'd
Her seasons change! behold! by slow degrees,
Stern winter turn'd into a ruder spring;
The ripen'd spring a milder summer glows;
Departing summer sheds Pomona's store;
And aged autumn brews the winter storm.

JOHN ARMSTRONG (170-9-1779)

The Snowdrop

Yes, punctual to the time, thou'rt here again,
As still thou art: though frost or rain may vary,
And icicles blockade the rockbirds' aery,
Or sluggish snow lie heavy on the plain,
Yet thou, sweet child of hoary January,
Art here to harbinger the laggard train
Of vernal flowers, a duteous missionary.
Nor cold can blight, nor fog thy pureness stain.
Beneath the dripping eaves, or on the slope
Of cottage garden, whether mark'd or no,
Thy meek head bends in undistinguish'd row.
Blessings upon thee, gentle bud of hope!
And Nature bless the spot where thou dost grow—
Young life emerging from thy kindred snow!

HARTLEY COLERIDGE (1796-1849)

To a Snowdrop

Lone flower, hemm'd in with snows, and white as they,
But hardier far, once more I see thee bend
Thy forehead, as if fearful to offend,
Like an unbidden guest. Though day by day,
Storms, sallying from the mountain-tops, waylay
The rising sun, and on the plains descend:
Yet art thou welcome, welcome as a friend
Whose zeal outruns his promise! Blue-eyed May
Shall soon behold this border thickly set
With bright jonquils, their odours lavishing
On the soft west wind and his frolic peers:
Nor will I then thy modest grace forget,
Chaste Snowdrop, venturous harbinger of Spring,
And pensive monitor of fleeting years!

WILLIAM WORDSWORTH (1770-1850)

Snowdrop

Could you understand
One who was wild as if he found a mine
Of golden guineas, when he noticed first
The soft green streaks in a Snowdrop's inner leaves?

ROBERT BUCHANAN (1841-1901)

Harbingers of Spring

For Snowdrops are the harbingers of Spring,
A sort of link between dumb life and light,
Freshness preserved amid all withering,
Bloom in the midst of grey and frosty blight,
Pale Stars that gladden Nature's dreary night!

CAROLINE ELIZABETH NORTON (1808-1877)

The Unloosening

Winter was weary. All his snows were failing—
Still from his stiff gray head he shook the rime
Upon the grasses, bushes and broad hedges,
But all was lost in the new touch of Time.

And the bright-globèd hedges were all ruddy,
As though warm sunset glowed perpetual.
The myriad swinging tassels of first hazel,
From purple to pale gold, were swinging all

In the soft wind, no more afraid of Winter.
Nor chaffinch, wren, nor lark was now afraid.
And Winter heard, or (ears too hard of hearing)
Snuffed the South-West that in his cold hair played.

And his hands trembled. Then with voice a-quaver
He called the East Wind, and the black East ran,
Roofing the sky with iron, and in the darkness
Winter crept out and chilled the earth again.

And while men slept the still pools were frozen,
Mosses were white, with ice the long grasses bowed;
The hawthorn buds and the greening honeysuckle
Froze, and the birds were dumb under that cloud.

And men and beasts were dulled, and children even
Less merry, under that low iron dome.
Early the patient rooks and starlings gathered;
Any warm narrow place for men was home.

And Winter laughed, but the third night grew weary,
And slept all heavy, till the East Wind thought him dead.
Then the returning South West in his nostrils
Breathed, and his snows melted. And his head

Uplifting, he saw all the laughing valley,
Heard the unloosened waters leaping down
Broadening over the meadows; saw the sun running
From hill to hill and glittering upon the town.

All day he stared. But his head drooped at evening,
Bent and slow he stumbled into the white
Cavern of a great chalk hill, hedged with tall bushes,
And in its darkness found a darker night

Among the broken cliff and falling water,
Freezing, or falling quietly everywhere;
Locked in a long, long sleep, his brain undreaming,
With only water moving anywhere.

Old men at night dreamed that they saw him going,
And looked, and dared not look, lest he should turn.
And young men felt the air beating on their bodies,
And the young women woke from dreams that burn.

And children going through the fields at morning
Saw the unloosened waters leaping down,
And broke the hazel boughs and wore the tassels
Above their eyes—a pale and shaking crown.

JOHN FREEMAN (1880-1929)

The Falling Night

Shepherds all, and maidens fair,
Fold your flocks up, for the air
'Gins to thicken, and the sun
Already his great course hath run.
See the dew-drops, how they kiss
Every little flower that is.
Hanging on their velvet heads,
Like a rope of crystal beads:
See the heavy clouds low falling,
And bright Hesperus down calling
The dead Night from under ground;
At whose rising, mists unsound,
Damps and vapours fly apace,
Hovering o'er the wanton face
Of these pastures, where they come,
Striking dead both bud and bloom:
Therefore, from such danger lock
Every one his lovèd flock;

And let your dogs lie loose without,
Lest the wolf come as a scout
From the mountain, and ere day,
Bear a lamb or kid away;
Or the crafty thievish fox
Break upon your simple flocks.
To secure yourselves from these,
Be not too secure in ease;
Let one eye his watches keep,
Whilst the other eye doth sleep;
So you shall good shepherds prove,
And for ever hold the love
Of our great god. Sweetest slumbers,
And soft silence fall in numbers
Of our eyelids! So, farewell!
Thus I end my evening's knell.

JOHN FLETCHER (1579-1625)

The New-Drop'd Lamb

Ah, gentle shepherd, thine the lot to tend,
Of all that feel distress, the most assailed,
Feeble, defenceless: lenient be thy care;
But spread around thy tend'rest diligence
In flow'ry spring-time, when the new-drop'd lamb,
Tott'ring with weakness by his mother's side,
Feels the fresh world about him; and each thorn,
Hillock, or furrow, trips his feeble feet.
O guard his meek sweet innocence from all
Th' innum'rous ills, that rush around his life;
Mark the quick kite, with beak and talons prone,
Circling the skies to snatch him from the plain;
Observe the lurking crows, beware the brake,
There the fly fox the careless minute waits;
Nor trust thy neighbour's dog, nor earth nor sky,
Thy bosom to a thousand cares divide.
Eurus oft slings his hail; the tardy fields
Pay not their promis'd food; and oft the dam
O'er her weak twins with empty udder mourns,
Or fails to guard, when the bold bird of prey
Alights, and hops in many turns around,
And tires her also turning: to her aid
Be nimble, and the weakest, in thine arms,
Gently convey to the warm cote, and oft,
Between the lark's note and the nightingale's,
His hungry bleating still with tepid milk;
In this office may thy children join;
And charitable habits learn in sport.

JOHN DYER (1699-1757)

The Primrose

Though storms may break the Primrose on its stalk,
Though frosts may blight the freshness of its bloom,
Yet Spring's awakening breath will woo the earth
To feed with kindliest dews its favourite flower,
That blooms in mossy banks and darksome glens,
Lighting the greenwood with its sunny smile,
Fear not then, Spirit, Death's disrobing hand.

PERCY BYSSHE SHELLEY (1792-1822)

Spring Flowers

The loveliest flowers the closest cling to earth,
And they first feel the sun; so violets blue,
So the soft star-like primrose drenched in dew,
The happiest of Spring's happy, fragrant birth,
To gentlest touches sweetest tones reply;—
Still humbleness with her low-breathed voice
Can steal o'er man's proud heart, and win his choice
From earth to heaven, with mightier witchery
Than eloquence or wisdom e'er could own.
Bloom on then in your shade, contented bloom,
Sweet flowers, nor deem yourselves to all unknown,
Heaven knows you, who one day for their altered doom
Shall thank you, taught by you to abase themselves and live.

JOHN KEBLE (1792-1866)

To an Early Primrose

Mild offspring of a dark and sullen sire!
Whose modest form, so delicately fine,
Was nursed in whirling storms,
And cradled in the winds.

Thee, when young spring first questioned Winter's sway,
And dared the sturdy blusterer to the fight,
Thee on this bank he threw
To mark his victory.

In this low vale, the promise of the year,
Serene, thou openest to the nipping gale,
Unnoticed and alone,
Thy tender elegance.

So virtue blooms, brought forth amid the storms
Of chill adversity; in some lone walk
Of life she rears her head,
Obscure and unobserved;

With every bleaching breeze that on her blows
Chastens her spotless purity of breast,
And hardens her to bear
Serene the ills of life.

HENRY KIRKE WHITE (1785-1806)

The Miracle

Come, sweetheart, listen, for I have a thing
Most wonderful to tell you—news of Spring.
Albeit Winter still is in the air
And the Earth troubled, and the branches bare,
Yet down the fields to-day I saw her pass—
The Spring—her feet went shining through the grass,
She touched the ragged hedgerows—I have seen
Her finger-prints, most delicately green;
And she has whispered to the crocus leaves,
And to the garrulous sparrows in the eaves.
Swiftly she passed and shyly, and her fair
Young face was hidden in her cloudy hair.
She would not stay, her season is not yet,
But she has reawakened, and has set
The sap of all the world astir, and rent
Once more the shadows of our discontent.
Triumphant news—a miracle I sing—
The everlasting miracle of Spring.

JOHN DRINKWATER (1882-1937)

24

Ploughing up the Pasture

Now up the pasture's slope the ploughed land laps
In folds that fall and crumble from the share
Rooks drop to the warm earth, hot leather creaks,
The sweat of labouring flesh steams in the air,
The flanks of beast are smooth with sun and toil,
The cropped turves that are worn with years of grazing
Turn inwards to the steel, and over the long
Acres of grassland stretch the ribs of soil.
No longer when in summer the clotted shadows
Fall from the crest of trees, will they stretch over
The lazy turf, but will shadow a new world
Of yellow acres, fret and stir of meadows,
Green barley, freckled silver by the wind.
And corn like a fresh sea across the world.

MARGARET STANLEY-WRENCH (20th century)

On Ploughing

The slow shuttle of husbandry
Has plodded up and down
Till folds of tilth are lying
In ripples of shining brown.

The slow thoughts of my ancestry
Are moving across my brain,
Turning today's deeds under,
Laying the old facts plain:

How my father strode at his furrowing,
My mother's father spun
And worked in the mills of weaving;
So the image of both is one . . .

The plough, horses and harnessing
Weaving slow lines of thread:
My grandfather and my father
Sweating for daily bread.

EVELYN D. BANGAY (20th century)

To the Small Celandine

Pansies, lilies, kingcups, daisies,
Let them live upon their praises;
Long as there's a sun that sets,
Primroses will have their glory;
Long as there are Violets,
They will have a place in story:
There's a flower that shall be mine,
'Tis the little Celandine.

Eyes of some men travel far
For the finding of a star;
Up and down the heavens they go,
Men that keep a mighty rout!
I'm as great as they, I trow,
Since the day I found thee out,
Little Flower!—I'll make a stir,
Like a great astronomer.

Modest, yet withal an Elf
Bold, and lavish of thyself;
Since we needs must first have met
I have seen thee, high and low,
Thirty years or more, and yet
'Twas a face I did not know;
Thou has now, go where I may,
Fifty greetings in a day.

Ere a leaf is on a bush,
In the time before the Thrush
Has a thought about her nest,
Thou wilt come with half a call,
Spreading out thy glossy breast
Like a careless Prodigal;
Telling tales about the sun,
When we've little warmth, or none.

Poets, vain men in their mood!
Travel with the multitude:
Never heed them; I aver
That they all are wanton wooers;
But the thrifty Cottager,
Who stirs little out of doors,
Joys to spy thee near her home;
Spring is coming, Thou are come!

Comfort have thou of thy merit,
Kindly unassuming Spirit!
Careless of thy neighbourhood,
Thou dost show thy pleasant face
On the moor, and in the wood,
In the lane—there's not a place,
Howsoever mean it be,
But 'tis good enough for thee.

Ill befall the yellow Flowers,
Children of the flaring hours!
Buttercups, that will be seen,
Whether we will see or no;
Others, too, of lofty mien;
They have done as worldlings do,
Taken praise that should be thine,
Little, humble Celandine!

Prophet of delight and mirth,
Ill-requited upon earth;
Herald of a mighty band,
Of a joyous train ensuing,
Serving at my heart's command,
Tasks that are no tasks renewing,
I will sing, as doth behove,
Hymns in praise of what I love!

WILLIAM WORDSWORTH (1770-1850)

April Birthday

When your birthday brings the world under your window
  And the song-thrush sings wet-throated in the dew
And aconite and primrose are unsticking the wrappers
  Of the package that has come today for you

      Lambs bounce out and stand astonished
      Puss willow pushes among bare branches
      Sooty hawthorns shiver into emerald

And a new air
      Nuzzles the sugary
      Buds of the chestnut. A groundswell and a stir
      Billows the silvered
      Violet silks
      Of the south—a tenderness
      Lifting through all the
      Gently-breasted
      Counties of England.

When the swallow snips the string that holds the world in
  And the ring-dove claps and nearly loops the loop
You just can't count everything that follows in the tumble
  Like a whole circus tumbling through a hoop

      Grass in a mesh of all flowers floundering
      Sizzling leaves and blossoms bombing
      Nestling hissing and groggy-legged insects

And the trees
      Stagger, they stronger
      Brace their boles and biceps under
      The load of gift. And the hills float
      Light as bubble glass
      On the smoke-blue evening

And rabbits are bobbing everywhere, and a thrush
Rings coolly in a far corner. A shiver of green
Strokes the darkening slope as the land
Begins her labour.

TED HUGHES (b 1930)

April

April, April,
Laugh thy girlish laughter;
Then, the moment after,
Weep thy girlish tears!
April, that mine ears
Like a lover greetest,
If I tell thee, sweetest,
All my hopes and fears,
April, April,
Laugh thy golden laughter,
But, the moment after,
Weep thy golden tears!

JOHN WILLIAM WATSON (1858-1935)

from *The Passionate Shepherd*

The fields are green, the spring grows on apace,
  And Nature's art begins to take the air;
Each herb her scent, each flower doth show her grace,
  And beauty braggeth of her bravest fair.
The lambs and rabbits sweetly run at base;
  The fowls do plume, and fishes fall to play;
The muses all have chose a sitting place
  To sing and play the shepherd's roundelay . . .
The little wren that never sung a note
  Is peeping now to prove how she can sing;
The nightingale hath set in tune her throat,
  And all the woods with little robins ring . . .
Love is abroad as naked as my nail,
  And little birds do flicker from their nests;
Diana sweet hath set aside her veil,
  And Phillis shows the beauty of her breasts.
O blessèd breasts, the beauty of the spring!
  O blessèd spring, that such a beauty shows!
Of highest trees the holly is the king,
  And of all flowers fair fall the queen, the rose!

NICHOLAS BRETON (?1553-?1625)

April Rise

If ever I saw blessing in the air
  I see it now in this still early day
Where lemon-green the vaporous morning drips
  Wet sunlight on the powder of my eye.

Blown bubble-film of blue, the sky wraps round
  Weeds of warm light whose every root and rod
Splutters with soapy green, and all the world
  Sweats with the bead of summer in its bud.

If ever I heard blessing it is there
  Where birds in trees that shoals and shadows are
Splash with their hidden wings and drops of sound
  Break on my ears their crests of throbbing air.

Pure in the haze the emerald sun dilates,
  The lips of sparrows milk the mossy stones,
While white as water by the lake a girl
  Swims her green hand among the gathered swans.

Now as the almond burns its smoking wick,
  Dropping small flames to light the candled grass;
Now, as my low blood scales its second chance,
  If ever world were blessed, now it is.

LAURIE LEE (b 1914)

Sudden Spring

Spring is sudden: it is her quality.
However carefully we watch for her,
However long delayed
The green in the winter'd hedge
The almond blossom
The piercing daffodil,
Like a lovely woman late for her appointment
She's suddenly here, taking us unawares,
So beautifully annihilating expectation
That we applaud her punctual arrival.

GERALD BULLETT (1893-1958)

By A Brook

Townsmen, or of the hamlet, young or old,
Whithersoever you may wander now,
Where'er you roam from, would you waste an hour,
Or sleep thro' one brief dream upon the grass, —
Pause here. The murmurs of the rivulet,
Rippling by cressy isles or bars of sand,
Are pleasant from the early Spring to when,
Full fields of barley shifting tearful lights
On growing spears, by fits the lady ash
With twinkling fingers sweeps her yellow keys.

ALFRED, LORD TENNYSON (1809-1892)

Winter Violets

Death-white azaleas watched beside my bed,
And tried to tell me tales of Southern lands;
But they in hothouse air were born and bred,
And they were gathered by a stranger's hands:
They were not sweet, they never have been free,
And all their pallid beauty had no voice for me.

And all I longed for was one common flower
Fed by soft mists and rainy English air,
A flower that knew the woods, the leafless bower
The wet, green moss, the hedges sharp and bare—
A flower that spoke my language, and could tell
Of all the woods and ways my heart remembers well.

Then came your violets—and at once I heard
The sparrows chatter on the dripping eaves
The full stream's babbling inarticulate word,
The plash of rain on big wet ivy leaves;
I saw the woods where thick the dead leaves lie,
And smelt the fresh earth's scent—the scent of memory.

The unleafed trees—the lichens green and grey,
The wide sad-coloured meadows, and the brown
Fields that sleep now, and dream of harvest day
Hiding their seeds like hopes in hearts pent down—
A thousand dreams, a thousand memories
Your violets' voices breathed in unheard melodies—

Unheard by all but me. I heard, I blessed
The little English, English-speaking things
For their sweet selves that laid my wish to rest,
For their sweet help that lent my dreaming wings,
And, most of all, for all the thoughts of you
Which make them smell more sweet than any other violets do.

EDITH NESBIT (1858-1924)

There's Many a Flower

The crocus flaunts its beauty in the sun;
The palm-fragrance greets us from afar;
The daisy shines as bright as any star;
The bluebell splashes woods while brake is dun;
With woof of green pale primrose-light is spun;
Black heaths are patched with coltsfoot-gold bizarre,
The snowdrop strews the fields like mottled spar;
The social daffodil hails every one—
All these do meet the rambler's careless eye
Unsought for, but the violet's sweet smile
Lies hidden, like a gem from vulgar pry,
'Mid lushest dew-beds, safe from all the glare
Of noons; first guerdon of all those who dare,
Mid Spring's caprices, wander mile on mile.

WILLIAM DOWSING (19th-early 20th century)

To Violets

Welcome, maids of honour,
You do bring
In the Spring
And wait upon her.

She has virgins many,
Fresh and fair
Yet you are
More sweet than any.

You're the maiden posies,
And so graced
To be placed
'Fore damask roses.

Yet, though thus respected
By and by
Ye do lie,
Poor girls, neglected.

ROBERT HERRICK (1591-1674)

A Violet
from *Hamlet*

A violet in the youth of primy nature,
Forward, not permanent, sweet, not lasting,
The perfume and suppliance of a minute;
No more.

WILLIAM SHAKESPEARE (1564-1616)

The Violet and the Rose

The violet in the wood, that's sweet to-day,
Is longer sweet than roses of red June;
Set me sweet violets along my way,
And bid the rose flower, but not too soon.
Ah violet, ah rose, why not the two?
Why bloom not all fair flowers the whole year through?
Why not the two, young violet, ripe rose?
Why dies one sweetness when another blows

AUGUSTA WEBSTER (1837-1894)

Nesting Birds

   Of birds, how each, according to her kind,
Proper materials for her nest can find;
And build a frame, which deepest thought in man
Would or amend, or imitate in vain?
How in small flights they know to try their young,
And teach the callow child her parent's song?
Why these frequent the plain, and those the wood?
Why ev'ry land has her specific brood?
Where the tall crane, or wandr'ing swallow goes,
Fearful of gathering winds and falling snows?
If into rocks or hollow trees they creep,
In temporary death confin'd to sleep.
Or conscious of the coming evil, fly
To milder regions, and a southern sky?

MATTHEW PRIOR (1664-1721)

The Voice of Spring

See the yellow catkins cover
All the slender willows over;
And on banks of mossy green
Star-like primroses are seen;
And, their clustering leaves below,
White and purple violets blow.

Hark! the new-born lambs are bleating,
And the cawing rooks are meeting
In the elms—a noisy crowd;
All the birds are singing loud,
And the first white butterfly
In the sunshine dances by.

Look around thee—look around!
Flowers in all the fields abound;
Every running stream is bright;
All the orchard trees are white;
And each small and waving shoot
Promises sweet flowers and fruit.

Turn thine eyes to earth and heaven
God for thee the spring has given,
Taught the birds their melodies,
Clothed the earth, and cleared the skies,
For thy pleasure or thy food—
Pour thy soul in gratitude!

MARY HOWITT (1799-1888)

from *A Year's Windfalls*

In the wind of windy March
  The catkins drop down,
Curly, caterpillar-like,
  Curious green and brown.
With concourse of nest-building birds
  And leaf-buds by the way,
We begin to think of flowers
  And life and nuts some day.

CHRISTINA ROSSETTI (1830-1894)

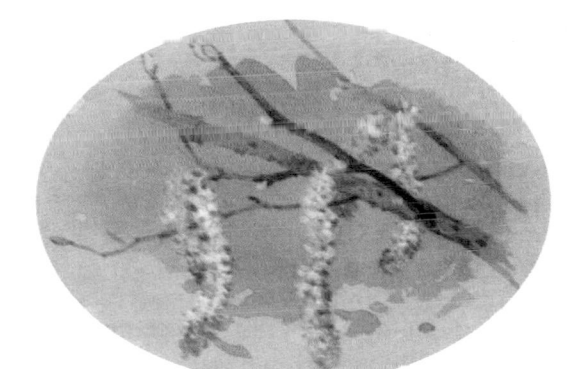

from *Misconceptions*

This is a spray the Bird clung to,
  Making it blossom with pleasure,
Ere the high tree-top she sprung to,
  Fit for her nest and her treasure.
  Oh, what a hope beyond measure
    Was the poor spray's, which the flying feet hung to,—
    So to be singled out, built in, and sung to!

ROBERT BROWNING (1812-1889)

from *'I Stood Tip-toe Upon a Little Hill'*

I gazed awhile, and felt as light, and free
As though the fanning wings of Mercury
Had played upon my heels: I was light-hearted,
And many pleasures to my vision started;
So I straightway began to pluck a posey
Of luxuries bright, milky, soft, and rosy.
A bush of May-flowers with the bees about them,
And let long grass grow round the roots to keep them
Moist, cool and green; and shade the violets
That they may bind the moss in leafy nets.

JOHN KEATS (1795-1821)

Description of Spring
Wherein Each Thing Renews, Save only the Lover

The sootë* season, that bud and bloom forth brings   *sweet
With green hath clad the hill, and eke the vale:
The nightingale with feathers now she sings;
The turtle to her mate hath told her tale;
Summer is come, for every spray now springs,
The hart hath hung his old head on the pale;
The buck in brake his winter coat he flings;
The fishes float with now repaired scale;
The adder all her slough away she slings;
The swift swallow pursueth the flies smale*;   *small
The busy bee her honey now she mings*;   *mingles
Winter is worn that was the flowers' bale.
  And thus I see among these pleasant things,
  Each care decays, and yet my sorrow springs.

HENRY HOWARD (?1517-1547)

The Flowering May-Thorn Tree

There stands the flowering May-thorn tree!
From through the veiling mist you see
The black and shadowy stem;
Smit by the sun, the mist in glee
Dissolves to lightsome jewel'ry,
Each blossom hath its gem.

SAMUEL TAYLOR COLERIDGE (1772-1834)

from *A Defence of English Spring*

. . . As yet but single,
The bluebells with the grasses mingle;
But soon their azure will be scrolled
Upon the primrose cloth-of-gold.
Yes, those are early lady-smocks,
The children crumple in their frocks,
And carry many a zig-zag mile,
O'er meadow, footpath, gate, and stile
To stick in pots and jugs to dress
Their cottages sills and lattices.

ROBERT BRIDGES (1844-1930)

from *The May Queen*

The honeysuckle round the porch has wov'n its bowers,
And by the meadow-trenches blow the faint sweet cuckoo-flowers;
And the wild marsh marigold shines like fire in swamps and hollows gray,
And I'm to be Queen o' the May, mother, I'm to be Queen o' the May.

ALFRED, LORD TENNYSON (1835-1913)

Fritillaries

But once I went through the lanes, over the sharp
Tilt of the little bridges; past the forge,
And heard the clang of anvil and iron,
And saw the founting sparks in the dusky forge,
And men outside with horses, gossiping.
So I came through that April England, moist
And green in its lush fields between the willows,
Foaming with cherry in the woods, and pale
With clouds of lady's-smock along the hedge,
Until I came to a gate and left the road
For the gentle fields that entice me, by the farms,
Wandering through the embroidered fields each one
So like its fellow; wandered through the gaps,
Past the mild cattle knee-deep in the brooks,
And wandered drowsing as the meadows drowsed
Under the pale wide heaven and slow clouds.
And then I came to a field where the springing grass
Was dulled by the hanging cups of fritillaries,
Sullen and foreign-looking, the snaky flower,
Scarfed in dull purple, like Egyptian girls
Camping among the furze, staining the waste
With foreign colour, sulky—dark and quaint,
Dangerous too, as a girl might sidle up,
An Egyptian girl, with an ancient snaring spell,
Throwing a net, soft round the limbs and heart,
Captivity soft and abhorrent, a close-meshed net,
—See the square web on the murrey flesh of the flower
Holding her captive close with her bare brown arms.

VICTORIA SACKVILLE-WEST (1892-1962)

Evening Orchard

All the beauty of the world
    Fading white,
All Spring's beauty, Summer's sweet,
    Wanton heat,
Wantoning 'neath the narrow bosom
    Of starry-flowered
Apple tree in late May blossom.

All the sweetness of the world
    Wanton White,
Paling cheek and fading heat,
    Wasting hue,
Stars the tree's thin breast with flower
    Like white stars
Shaken from sky-arching bower

All the beauty, all the sweetness
    Staying white:
The ruddy gone, the brief pink gone,
    And bridal light
Shining yet while grasses flush
    With evening's gold,
And Eden's is the orchard thrush.

JOHN FREEMAN (1880-1929)

from *A Year's Windfalls*

With the gusts of April
    Rich fruit-tree blossom fall,
On the hedged-in orchard-green,
    From the southern wall.
Apple-trees and pear-trees
    Shed petals white or pink,
Plum-trees and peach-trees;
    While sharp showers sink and sink.

CHRISTINA ROSSETTI (1830-1894)

from *The Seasons*

...And see the country, far diffus'd around,
One boundless blush, one white impurpled shower
Of mingled blossoms; where the raptur'd eye
Hurries from joy to joy.

JAMES THOMSON (1700-1748)

from *Song of the Bluebells*

Sweet bluebells we,
Mid flowers of the lea,
The likest in hue to heaven
Our bonnets so blue
Are tinged with the dew
That drops from the sky at even.

Our bloom more sweet
Than dark violet,
Or tulip's purple stain,
At every return
Of the dew-breathing morn,
Grows brighter and brighter again.

GEORGE DARLEY (1795-1846)

from *Bluebells*

Tonight from deeps of loneliness I wake in wistful wonder
To a sudden sense of brightness, an immanence of blue—
O are there bluebells swaying in the shadowy coppice yonder,
Shriven with the dawning and the dew?

LUCIA C. MARKHAM (late 19th-early 20th century)

Bluebells

'One day, one day, I'll climb that distant hill
    And pick the bluebells there!'
So dreamed the child who lived beside the rill
And breathed the lowland air.
    'One day, one day when I am old I'll go
And climb the mountain where the bluebells blow.'

'One day! one day!' The child was now a maid,
    A girl with laughing look;
She and her lover sought the valley-glade
Where sang the silver brook.
    'One day,' she said, 'love, you and I will go
And reach that far hill where the bluebells blow!'

Years passed. A woman now with wearier eyes
    Gazed towards that sunlit hill.
Tall children clustered round her. How time flies!
The bluebells blossomed still.
    She'll never gather them! All dreams fade so.
We live and die, and still the bluebells blow.

GEORGE BARLOW (19th century)

46

On a Bed of Forget-me-nots

I love its growth at large and free
By untrod path and unlopped tree,
Or nodding by the unpruned hedge,
Or on the water's dangerous edge
Where flags and meadowsweet blow rank
With rushes on the quaking bank.

CHRISTINA ROSSETTI (1830-1894)

from *The Keepsake*

Nor can I find, amid my lonely walk
By rivulet, or spring, or wet roadside,
That blue and bright-eyed floweret of the brook,
Hope's gentle gem, the sweet Forget-me-not.

SAMUEL TAYLOR COLERIDGE (1772-1834)

from *On May Morning*

The little blue Forget-me-not
Comes too on friendship's gentle plea,
Spring's messenger in every spot,
Smiling on all,— 'Remember me!'

JOHN CLARE (1793-1864)

from *Auguries of May*

The sedge-wren tells her note,
  Dim larks in ether float,
The uprolled clouds sustain their pageant dome.
  In velvet, sunshine-fed,
  Spires up the bulrush head,
Where rock the wild swans in their reedy home.

  The lily pale and wan
  Puts all her glories on:
Her silver mantle and her golden crest.
  The humbler violets stand
  Her ladies at command,
As she attires in lawn her ivory breast.

JOHN BYRNE LEICESTER WARREN (1835-1895)

from *The Gift: for C.M.P.*

The mill-wheel, cheerful drudge, would roll
And splash and drum, but the bright-eyed vole
Would never care for him, would swim
Across his racing waves, and slim
Sharp dace would watch in the quickest gush,
And forget-me-not and flag and rush
Would take up quarters there, boom as he might.

EDMUND BLUNDEN (1896-1974)

49

The Mood of May

When May is in his prime, then may each heart rejoice:
When May bedecks each branch with green, each bird strains forth his voice.
The lively sap creeps up into the blooming thorn;
The flowers, which cold in prison kept, now laugh the frost to scorn.
All nature's imps triumph whiles joyful May doth last;
When May is gone, of all the year the pleasant time is past.

May makes the cheerful hue, May breeds and brings new blood;
May marcheth throughout every limb, May makes the merry mood.
May pricketh tender hearts their warbling notes to tune:
Full strange it is, yet some we see do make their May in June.
Thus things are strangely wrought whiles joyful May doth last;
Take May in time, when May is gone the pleasant time is past.

All ye that live on earth, and have your May at will,
Rejoice in May, as I do now, and use your May with skill.
Use May while that you may, for May hath but his time,
When all the fruit is gone, it is too late the tree to climb.
Your liking and your lust is fresh whiles May doth last;
When May is gone, of all the year the pleasant time is past.

RICHARD EDWARDS (1524-1566)

At the Close of Spring

The garlands fade that Spring so lately wove,
  Each simple flower which she had nursed in dew,
Anemones that spangled every grove,
  The primrose wan, and hare-bell mildly blue.
No more shall violets linger in the dell,
  Or purple orchis variegate the plain,
Till Spring again shall call forth every bell,
  And dress with humid hands her wreaths again.
Ah! poor humanity! so frail, so fair,
  Are the fond visions of thy early day,
Till tyrant passion and corrosive care
  Bid all thy fairy colours fade away!
Another May new buds and flowers shall bring,
Ah! why has happiness no second Spring?

CHARLOTTE SMITH (1749-1806)

Early June

Spring's over, over. The gold meadows tarnish,
The gold dims, heavy-leaved hedges darken,
The primal light diminishes.
I look, look back, and hearken
Now but to faint and ever fainter echoes.
Summer lays siege, and Spring's brief fire finishes.

Never was such a glory as this Spring glory,
Never a cloudy navy of such brightness
Moving all day to nights serener.
But I, who shared that lightness,
Feel already the season's weight more sombre,
Already the leaves falling, and the brave boughs grown leaner.

LAURENCE BINYON (1869-1943)

To the Dandelion

Dear common flower, that grow'st beside the way,
Fringing the dusty road with harmless gold,
First pledge of blithesome May,
Which children pluck, and full of pride uphold,
High-hearted buccaneers, o'erjoyed that they
An Eldorado in the grass have found,
Which not the rich earth's ample round
May match in wealth! Thou art more dear to me
Than all the prouder summer blooms may be.

Gold such as thine ne'er drew the Spanish prow
Through the primeval hush of Indian seas,
Nor wrinkled the lean brow
Of age, to rob the lover's heart of ease;
'Tis the Spring's largess, which she scatters now
To rich and poor alike, with lavish hand,
Though most hearts never understand
To take it at God's value, but pass by
The offered wealth with unrewarded eye.

Thou art my tropics and mine Italy;
To look at thee unlocks a warmer clime;
The eyes thou givest me
Are in the heart, and heed not space or time:
Not in mid June the golden-cuirassed bee
Feels a more summer-like warm ravishment
In the white lily's breezy tent,
His fragrant Sybaris, than I when first
From the dark green thy yellow circles burst.

Then think I of deep shadows on the grass,
Of meadows where in sun the cattle graze,
Where, as the breezes pass,
The gleaming rushes lean a thousand ways,
Of leaves that slumber in a cloudy mass,
Or whiten in the wind, of waters blue
That from the distance sparkle through
Some woodland gap, and of a sky above,
Where one white cloud like a stray lamb doth move.

My childhood's earliest thoughts are linked with thee;
The sight of thee calls back the robin's song,
Who, from the dark old tree
Beside the door, sang clearly all day long;
And I, secure in childish piety,
Listened as if I heard an angel sing
With news from heaven, which he could bring
Fresh every day to my untainted ears
When birds and flowers and I were happy peers.

How like a prodigal doth nature seem,
When thou, for all thy gold, so common art!
Thou teachest me to deem
More sacredly of every human heart,
Since each reflects in joy its scanty gleam
Of heaven, and could some wondrous secret show,
Did we but pay the love we owe,
And with a child's undoubting wisdom look
On all these living pages of God's book.

JAMES RUSSELL LOWELL (1819-1891)

55

A Summer's Day

My Guardian, bear me on thy downy wing
To some cool shade, where infant flowers spring,
Where on the trees sweet honeysuckles blow,
And ruddy daisies paint the ground below:
Where the shrill linnet charms the solemn shade,
And zephyrs pant along the cooler glade,
Or shake the bull-rush by a river-side,
While the gay sun-beams sparkle on the tide:
Oh! for some grot, whose rustic sides declare
Ease, and not splendour, was the builder's care;
Where roses feed their unaffected charms,
And the curl'd vine extends her clasping arms:
Where happy silence lulls the quiet soul,
And makes it calm as summer waters roll.
Here let me learn to check each growing ill,
And bring to reason disobedient will;
To watch this incoherent breast, and find
What fav'rite passions rule the giddy mind.
Here no reproaches grate the wounded ear;
We see delighted, and transported hear:
While the glad warblers wanton round the trees,
And the still waters catch the dying breeze.
Grief waits without, and melancholy gloom;
Come, chearful hope, and fill the vacant room;
Come, ev'ry thought, which virtue gave to please;
Come, smiling health! with thy companion, ease.
Let these, and all that virtue's self attends,
Bless the still hours of my gentle friends.
Peace to my foes, if any such there be,
And gracious heaven give kind repose to me.

MARY LEAPOR (1722-1746)

from *Ballad of Dowsabel*

This maiden in a morn betime,
Went forth when May was in the prime,
  To get sweet setywall,
The honey-suckle, the harlock
The lily, and the lady-smock,
  To deck her summer hall.

MICHAEL DRAYTON (1563-1631)

Honeysuckle
from *Much Ado about Nothing*

And bid her steal into the pleached bower,
Where honeysuckles, ripen'd by the sun,
Forbid the sun to enter, like favourites,
Made proud by princes . . .

WILLIAM SHAKESPEARE (1564-1616)

from *Gareth and Lynette*

Good Lord, how sweetly smells the honeysuckle
In the hush'd night, as if the world were one
Of utter peace, and love, and gentleness.

ALFRED, LORD TENNYSON (1809-1892)

To The Daisy

With little here to do or see
Of things that in the great world be,
Daisy! again I talk to thee,
   For thou art worthy,
Thou unassuming Common-place
Of Nature, with that homely face,
And yet with something of a grace
   Which love makes for thee!

Oft on the dappled turf at ease
I sit, and play with similes,
Loose types of things, through all degrees,
   Thoughts of thy raising:
And may a fond and idle name
I give to thee, for praise or blame,
As is the humour of the game,
   While I am gazing.

A nun demure of lowly port;
Or sprightly maiden, of Love's court,
In thy simplicity the sport
   Of all temptations;
A queen in crown of rubies drest;
A starveling in a scanty vest;
Are all, as seems to suit thee best,
   Thy appellations.

A little Cyclops with one eye
Staring to threaten and defy,
That thought comes next—and instantly
   The freak is over,
The shape will vanish—and behold
A silver shield with boss of gold,
That spreads itself, some faery bold
   In fight to cover!

I see thee glittering from afar—
And then thou art a pretty star;
Not quite so fair as many are
   In heaven above thee!
Yet like a star, with glittering crest,
Self-poised in air thou seem'st to rest;—
May peace come never to his nest,
   Who shall reprove thee!

Bright flower! for by that name at last,
When all my reveries are past;
I call thee, and to that cleave fast,
   Sweet silent creature!
Thou breath'st with me in sun and air,
Do thou, as thou art wont, repair
   My heart with gladness, and a share
   Of thy meek nature!

WILLIAM WORDSWORTH (1770-1850)

from *The Legend of Good Women*

The long day I shape me for to abide,
For nothing else, and I shall not lie,
But for to look upon the Daisie,
That well by reason men it call may
The Daisie, or else the Eye of the Day,
The empress and flowre of flowres all.

GEOFFREY CHAUCER (c.1342-1400)

from *The Eternity of Nature*

Trampled under foot,
The daisy lives and strikes its root
Into the lap of time; centuries may come
And pass away into the silent tomb,
And still the child, hid in the womb of time,
Shall smile and pluck them, when this simple rhyme
Shall be forgotten, like a churchyard and stone,
Or lingering lie, unnoticed and alone.

JOHN CLARE (1793-1864)

The Bee

The pedigree of honey
Does not concern the bee
A clover, any time, to him
Is aristocracy.

EMILY DICKINSON (1830-1886)

from *The Seasons*

Here their delicious task the fervent bees
In swarming millions tend; around, athwart,
Through the soft air the busy nations fly,
Cling to the bud, and with inserted tube
Suck its pure essence, its aetherieal soul;
And oft, with bolder wing, they soaring dare
The purple heath, or where the wild thyme grows,
And yellow load them with the luscious spoil.

JAMES THOMSON (1700-1748)

A Procession

Marvellous wings filled the morning:
The bourdon bee from grass
To grass heaved his brown sacks;
The butterfly battled with air,
Adorning her wings with light.
Beetles with armoured backs
Flashed steel and bronze so bright
That a king, it seemed, must pass
For the hordes of the orchard to stare,
Raise huzzah and buzz
With rustic gossamer wing,
Their acclamation thus
Catching sunshine, noon-sound,
Hay-height above the ground,
Though none quite glimpsed the king.

RICHARD CHURCH (1893-1972)

from *The Months*

Heavy is the green of the fields, heavy the trees
With foliage hang, drowsy the hum of bees
In the thund'rous air: the crowded scents lie low:
Thro' tangle of weeds the river runneth slow.

ROBERT BRIDGES (1844-1930)

The Careful Insect
from *Rural Sports*

The careful Insect 'midst his works I view
Now from the Flow'rs exhaust the fragrant Dew,
With golden Treasures load his little Thighs,
And steer his airy Journey through the Skies:
With liquid Sweets the waxen Cells distend,
While some 'gainst Hostile Drones their cave defend
Others with sweets the waxen cells distend:
Each in the Toil a proper Station bears,
And in the little Bulk a mighty Soul appears.

JOHN GAY (1685-1732)

The Flitting

I've left my own old home of homes,
Green fields and every pleasant place;
The summer like a stranger comes,
I pause and hardly know her face.
I miss the hazel's happy green,
The blue bell's quiet hanging blooms,
Where envy's sneer was never seen,
Where staring malice never comes.

I miss the heath, its yellow furze,
Molehills and rabbit tracks that lead
Through beesom, ling, and teazel burrs
That spread a wilderness indeed;
The woodland oaks and all below
That their white powdered branches shield;
The mossy paths: the very crow
Croaked music in my native fields.

I sit me in my corner chair
That seems to feel itself at home,
And hear bird music here and there
From hawthorn hedge and orchard come.
I hear, but all is strange and new:
I sat on my old bench in June,
The sailing puddock's shrill 'peelew'
On Royce Wood seemed a sweeter tune.

I walk adown the narrow lane,
The nightingale is singing now,
But like to me she seems at loss
For Royce Wood and its shielding bough.
I lean upon the window sill,
The bees and summer happy seem;
Green, sunny green they shine, but still
My heart goes far away to dream

Of happiness, and thoughts arise
With home-bred pictures many a one,
Green lanes that shut out burning skies
And old crook'd stiles to rest upon;
Above them hangs the maple tree,
Below grass swells a velvet hill,
And little footpaths sweet to see
Go seeking sweeter places still.

JOHN CLARE (1793-1864)

from *The Prelude*

Through quaint obliquities I might pursue
These cravings; when the foxglove, one by one,
Upwards through every stage of the tall stem,
Had shed beside the public way its bells,
And stood of all dismantled, save the last
Left at the tapering ladder's top, that seemed
To bend as doth a slender blade of grass
Tipped with a rain-drop . . .

WILLIAM WORDSWORTH (1770-1850)

from *Thyrsis*

So, some tempestuous morn in early June,
When the year's primal burst of bloom is o'er,
Before the roses and the longest day—
When garden-walks, and all the grassy floor,
With blossoms, red and white, of fallen May
And chestnut-flowers are strewn—
So have I heard the cuckoo's parting cry
From the wet field, through the vext garden-trees,
Come with the volleying rain and tossing breeze:
The bloom is gone, and with the bloom go I.

Too quick despairer, wherefore wilt thou go?
Soon will the high Midsummer pomps come on,
Soon will the musk carnations break and swell,
Soon shall we have gold dusted snapdragon,
Sweet-William with its homely cottage-smell,
And stocks in fragrant blow;
Roses that down the alleys shine afar,
And open, jasmine-muffled lattices,
And groups under the dreaming garden-trees,
And the full moon, and the white evening-star.

MATTHEW ARNOLD (1822-1888)

Summer

Winter is cold-hearted,
  Spring is yea and nay,
Autumn is a weather-cock
  Blown every way:
Summer days for me,
When every leaf is on its tree:

When Robin's not a beggar,
  And Jenny Wren's a bride,
And larks hang singing, singing, singing,
  Over the wheat fields wide,
  And anchored lilies ride,
And the pendulum spider
  Swings from side to side,

And blue black beetles transact business,
  And gnats fly in a host,
And furry caterpillars hasten
  That no time be lost,
And moths grow fat and thrive,
And ladybirds arrive.

Before green apples blush,
  Before green nuts embrown,
Why, one day in the country
  Is worth a month in town:
  Is worth a day and a year
Of the dusty, musty, lag-last fashion
  That days drone elsewhere.

CHRISTINA ROSSETTI (1830-1894)

from *The Month of June*

It is the month of June
The month of leaves and roses,
When pleasant sights salute the eyes
And pleasant scents the noses.

NATHANIEL PARKER WILLIS (1806-1894)

To—

The roses make the world so sweet,
The bees, the birds have such a tune,
There's such a light and such a heat
And such a joy in June.

GEORGE MACDONALD (1824-1905)

Summer Night

I cannot see what flowers are at my feet,
   Nor what soft incense hangs upon the boughs,
But in embalmed darkness, guess each sweet
   Wherewith the seasonable month endows
The grass, the thicket, and the fruit-tree wild;
   White hawthorn and the pastoral eglantine;
Fast-fasting violets cover'd up in leaves;
   And mid-May's eldest child,
The coming musk-rose, full of dewy wine,
   The murmurous haunt of flies on summer eves.

JOHN KEATS (1795-1821)

from *The Vision of Sir Launfal*

And what is so rare as a day in June?
Then, if ever, come perfect days;
Then Heaven tries earth if it be in tune,
And over it softly her warm ear lays;
Whether we look, or whether we listen,
We hear life murmur, or see it glisten;
Every clod feels a stir of might,
An instinct within it that reaches and towers,
And, groping blindly above it for light,
Climbs to a soul in grass and flowers;
The flush of life may well be seen
Thrilling back over hills and valleys;
The cowslip startles in meadows green,
The buttercup catches the sun in its chalice,
And there's never a leaf nor a blade too mean
To be some happy creature's palace;
The little bird sits at his door in the sun,
Atilt like a blossom among the leaves,
And lets his illumined being o'errun
With the deluge of summer it receives;
His mate feels the eggs beneath her wings,
And the heart in her dumb breast flutters and sings.
He sings to the wide world, and she to her nest,—
In the nice ear of Nature which song is the best?

JAMES RUSSELL LOWELL (1819-1891)

A Green Cornfield

'And singing still dost soar and soaring ever singest.'

The earth was green, the sky was blue:
   I saw and heard one sunny morn
A skylark hang between the two,
   A singing speck above the corn;

A stage below, in gay accord,
   White butterflies danced on the wing,
And still the singing skylark soared,
   And silent sank and soared to sing.

The cornfield stretched a tender green
   To right and left beside my walks;
I knew he had a nest unseen
   Somewhere among the million stalks.

And as I paused to hear his song
   While swift the sunny moments slid,
Perhaps his mate sat listening long,
   And listened longer than I did.

CHRISTINA ROSSETTI (1830-1894)

The Lark

The lark, that shuns on lofty boughs to build
Her annual nest, lies silent in the field.
But if the promise of a cloudless day,
Aurora smiling, bids her rise and play,
Then straight she shews, 'twas not for want of voice,
Or power to climb, she made so low a choice:
Singing she mounts, her airy wings are stretch'd
T'wards heav'n, as if from heav'n her notes she fetch'd.

EDMUND WALLER (1606-1687)

The Heaven-Soaring Lark

The heaven-soaring lark, its rapture spent
On morning's quest
Drops down again, soul satisfied, content
Unto the nest.

O singing soul, chafe not, that by earth's chain
Thou seemest bound!—
The sky's true messenger did ne'er disdain
The lowly ground.

MARY ELEANOR ROBERTS (late 19th-early 20th century)

Harebell and Pansy

O'er the round throat her little head
Its gay delight upbuoys:
A harebell in the breeze of June
Hath such melodious poise;
And chiming with her heart, my heart
Is only hers and joy's.

But my heart takes a deeper thrill,
Her cheek a rarer bloom,
When the sad mood comes rich as glow
Of pansies dipped in gloom.
By some far shore she wanders—where?
And her eyes fill—for whom?

LAURENCE BINYON (1869-1943)

The Harebell

. . .On the windy hills
Lo, the little harebell leans
On the spire-grass that it queens
With bonnet blue.

GEORGE MEREDITH (1828-1909)

from *The Prelude*

Or, not less pleased, lay on some turret's head,
Catching from tufts of grass and hare-bell flowers
Their faintest whisper to the passing breeze,
Given out, while mid-day heat oppressed the plains.

WILLIAM WORDSWORTH (1770-1850)

from *The Seasons*

Now, flaming up the heav'ns, the potent sun
Melts into limpid air the high rais'd clouds,
In partly colour'd bands; till wide unveil'd,
The face of nature shines, from where earth seems
Far stretched around, to meet the bending sphere.

JAMES THOMSON (1700-1748)

The Sun

The sun that rolls his beamy orbs on high,
Pride of the world and glory of the sky;
Illustrious in his course, in bright array
Marches along the heavenes, and scatters day
O'er earth, and o'er the main, and the etherial way.
He in the morn renews his radiant round,
And warms the fragrant bosom of the ground;
But e'er the noon of day, in fiery gleams
He darts the glory of his blazing beams;
Beneath the burnings of his sultry ray,
Earth to her centre pierced, admits the day;
Huge vales expand, where rivers roll'd before,
And lessen'd seas contract within their shore.

WILLIAM BROOME (1689-1745)

The Sun

The sun, like a golden knot on high,
Gathers the glories of the sky,
And binds them into a shining tent,
Roofing the world with the firmament.
And through the pavilion the rich winds blow,
And through the pavilion the waters go,
And the buds for joy, and the trees for prayer,
Bowing their heads in the sunny air . . .

GEORGE MACDONALD (1824-1905)

77

The Brook

Seated once by a brook, watching a child
Chiefly that paddled, I was thus beguiled.
Mellow the blackbird sang and sharp the thrush
Not far off in the oak and hazel bush,
Unseen. There was a scent like honeycomb
From mugwort dull. And down upon the dome
Of the stone the cart-horse kicks against so oft
A butterfly alighted. From aloft
He took the heat of the sun, and from below,
On the hot stone he perched contented so,
As if never a cart would pass again
That way; as if I were the last of men
And he the first of insects to have earth
And sun together and to know their worth.
I was divided between him and the gleam,
The motion, and the voices, of the stream,
The waters running frizzled over gravel,
That never vanish and for ever travel.
A grey flycatcher silent on a fence
And I sat as if we had been there since
The horseman and the horse lying beneath
The fir-tree-covered barrow on the heath,
The horseman and the horse with silver shoes,
Galloped the downs last. All that I could lose
I lost. And then the child's voice raised the dead.
'No one's been here before' was what she said
And what I felt, yet never should have found
A word for, while I gathered sight and sound.

EDWARD THOMAS (1878-1917)

Sky

. . . Look overhead
How air is azurëd;
O how! nay do but stand
Where you can lift your hand
Skywards: rich, rich it laps
Round the four fingergaps.
Yet such a sapphire-shot,
Charged, steepëd sky will not
Stain light. Yea, mark you this:
It does no prejudice.
The glass-blue days are those
When every colour glows,
Each shape and shadow shows.
Blue be it: this blue heaven
The seven or seven times seven
Hued sunbeam will transmit
Perfect, not alter it.
Or if there does some soft,
On things aloof, aloft,
Bloom breathe, that one breath more
Earth is the fairer for . . .

GERARD MANLEY HOPKINS (1844-1889)

from *Thunder on the Downs*

Wide earth, wide heaven, and in the summer air
Silence! The summit of the Down is bare
Between the climbing crests of wood; but those
Great sea-winds, wont, when the wet South-West blows,
To rock tall beeches and strong oaks aloud
And strew torn leaves upon the streaming cloud,
To-day are idle, slumbering far aloof.
Under the solemn height and gorgeous roof
Of cloud-built sky, all earth is indolent.
Wandering hum of bees and thymy scent
Of the short turf enrich pure loneliness;
Scarcely an airy topmost-twining tress
Of bryony quivers where the thorn it wreathes;
Hot fragrance from the honeysuckle breathes,
And sweet the rose floats on the arching briar's
Green fountains sprayed with delicate frail fires.

For clumps of thicket, dark beneath the blaze
Of the high westering sun, beset the ways
Of smooth grass narrowing where the slope runs steep
Down to green woods, and glowing shadows keep
A freshness round the mossy roots, and cool
The light that sleeps as in a chequered pool
Of golden air. O woods, I love you well,
I love the flowers you hide, your ferny smell;
But here is sweeter solitude, for here
My heart breathes heavenly space; the sky is near
To thought, with heights that fathomlessly glow;
And the eye wanders the wide land below.

And this is England! June's undarkened green
Gleams on far woods; and in the vales between
Gray hamlets, older than the trees that shade
Their ripening meadows, are in quiet laid,
Themselves a part of the warm, fruitful ground. . .

83

Poppies in the Wheat

Along Ancona's hills the shimmering heat,
A tropic tide of air, with ebb and flow
Bathes all the fields of wheat until they glow
Like flashing seas of green, which toss and beat
Around the vines. The poppies lithe and fleet
Seem running, fiery torchmen, to and fro
To mark the shore. The farmer does not know
That they are there. He walks with heavy feet,
Counting the bread and wine by autumn's gain,
But I—I smile to think that days remain
Perhaps to me in which, though bread be sweet
No more, and red wine warm my blood in vain,
I shall be glad remembering how the fleet,
Lithe poppies ran like torchmen with the wheat.

HELEN HUNT JACKSON (1830-1885)

from *The Poppy*

Summer set lip to earth's bosom bare,
And left the flushed print in a poppy there:
Like a yawn of fire from the grass it came,
And the fanning wind puffed it to flapping flame.

With burnt mouth, red like a lion's, it drank
The blood of the sun as he slaughtered sank,
And dipped its cup in the purpurate shine
When the eastern conduits ran with wine.

Till it grew lethargied with fierce bliss,
And hot as a swinkèd gipsy is,
And drowsed in sleepy savageries,
With mouth wide a-pout for a sultry kiss.

FRANCIS THOMPSON (1859-1907)

The Sheaves

Where long the shadows of the wind had rolled,
Green wheat was yielding to the change assigned,
And as by some vast magic undivined
The world was turning slowly into gold.
Like nothing that was ever bought or sold
It waited there, the body and the mind;
And with a mighty meaning of a kind
That tells the more the more it is not told.

So in a land where all days are not fair,
Fair days went on till on another day
A thousand golden sheaves were lying there,
Shining and still, but not for long to stay—
As if a thousand girls with golden hair
Might rise from where they slept and go away.

EDWIN ARLINGTON ROBINSON (1869-1935)

September 1815

While not a leaf seems faded; while the fields,
With ripening harvest prodigally fair,
In brightest sunshine bask; this nipping air,
Sent from some distant clime where Winter wields
His icy scimitar, a foretaste yields
Of bitter change, and bids the flowers beware;
And whispers to the silent birds, 'Prepare
Against the threatening foe your trustiest shields.'
To Nature's tuneful quire, this rustling dry
Through leaves yet green, and yon crystalline sky,
Announce a season potent to renew,
'Mid frost and snow, the instinctive joys of song,
And nobler cares than listless summer knew.

WILLIAM WORDSWORTH (1770-1850)

Vanished Summers

Vanished Summers, passed and gone,
Here find resurrection.—
Each crowned corn-head closely filled,
Packed and pressed with suns distilled
Into lively sap which throws
Rays of sunlight as it grows.—
These enchanted, waving tall
Golden ears contain them all:
All the long delightful days,
When June met us face to face;
Light and laughing grace re-born
In great fields of upright corn.—
Earth's tremendous charity
Full-accomplished here we see
Who gives us for familiar food
The lovely lilt of July's mood.—
One minute, brown husk contains
Summer's shadow, Autumn rains,
Spring's delicious wayward green,
Even Winter's pallid, lean
Blood of mingled frost and snows
Virtue on our sheaves bestows.
So to give us daily bread
The very sky's transfiguréd.

MARGARET SACKVILLE (1881-1963)

Hay-Making

from *The Seasons*

Now swarms the village o'er the jovial mead:
The rustic youth, brown with meridian toil,
Healthful and strong, full as the summer rose;
Blown by prevailing suns, the ruddy maid,
Half-naked, swelling on the sight, and all
Her kindled graces burning o'er her cheek.
Ev'n stooping age is here; and infant hands
Trail the long rake, or, with the fragrant load,
O'er-charged, amid the kind oppression roll.
Wide flies the tedded grain; all in a row,
Advancing broad, or wheeling round the field,
They spread the breathing harvest to the sun,
That throws refreshful round a rural smell:
Or, as they rake the green appearing ground,
And drive the dusky wave along the mead,
The russet hay-cock rises thick behind,
In order gay; while heard from dale to dale,
Working the breeze resounds the blended voice
Of happy labour, love, and social glee.

JAMES THOMSON (1700-1748)

from *Seasons*

Oh the shouting Harvest-weeks!
  Mother Earth grown fat with sheaves;
Thrifty gleaner finds who seeks;
  Russet-golden pomp of leaves
Crowns the woods, to fall at length;
  Bracing winds are felt to stir,
Ocean gathers up her strength,
  Beasts renew their dwindled fur.

CHRISTINA ROSSETTI (1830-1894)

from *The Seasons*

The sun has lost his rage; his downward orb
Shoots nothing now, but animating warmth,
And vital lustre; that, with various ray,
Lights up the clouds, those beauteous robes of heav'n,
Incessant roll'd into romantic shapes,
The dream of waking fancy! broad below
Cover'd with ripening fruits, and swelling fast
Into the perfect year, the pregnant earth
And all her tribes rejoice.

JAMES THOMSON (1700-1748)

My Father's Fields

Not all the sights your boasted garden yields,
Are half so lovely as my father's fields,
Where large increase has bless'd the fruitful plain,
And we with joy behold the swelling grain!
Whose heavy ears, toward the earth reclin'd,
Wave, nod, and tremble to the whisking wind.

MARY LEAPOR (1722-1746)

A Day Born of the Gentle South

After dark vapours have oppress'd our plains
For a long dreary season, comes a day
Born of the gentle South, and clears away
From the sick heavens all unseemly stains.
The anxious month, relievèd of its pains,
Takes as a long-lost right the feel of May;
The eyelids with the passing coolness play
Make rose-leaves with the drip of Summer rains.
The calmest thoughts come round us; as of leaves
Budding—fruit ripening in stillness—Autumn suns,
Smiling at eve upon the quiet sheaves—. . .
Sweet Sappho's cheek—a smiling infant's breath—
The gradual sand that through an hour-glass runs—
A woodland ruvulet—a Poet's death.

JOHN KEATS (1795-1821)

October
from *The Months*

On frosty morns with the woods aflame, down, down
The golden spoils fall thick from the chestnut crown.
May Autumn in tranquil glory her riches spend,
With mellow apples her orchard-branches bend.

ROBERT BRIDGES (1844-1930)

from *The Lotos-Eaters*

Lo! sweetened with the summer light,
The full-juiced apple, waxing over-mellow,
Drops in a silent autumn night.
All its allotted length of days,
The flower ripens in its place,
Ripens and fades, and falls, and hath no toil,
Fast rooted in the fruitful soil.

ALFRED, LORD TENNYSON (1809-1892)

from *A Year's Windfalls*

In brisk wind of September
  The heavy-headed fruits
Shake upon their bending boughs
  And drop from the shoots;
Some glow golden in the sun,
  Some show green and streaked,
Some set forth a purple bloom,
  Some blush rosy-cheeked.

CHRISTINA ROSSETTI (1830-1894)

Martins: September

At secret daybreak they had met—
Chill mist beneath the welling light
Screening the marshes green and wet—
An ardent legion wild for flight.

Each preened and sleeked an arrowlike wing,
Then eager throats with lapsing cries
Praising whatever fate might bring—
Cold wave, or Africa's paradise.

Unventured, trackless leagues of air,
England's sweet summer narrowing on
Her lovely pastures; nought they care—
Only this ardour to be gone.

A tiny, elflike, ecstatic host . . .
And I neath them, on the highway's crust,
Like some small mute belated ghost,
A sparrow pecking in the dust.

WALTER DE LA MARE (1873-1956)

Bright Autumn

I met not any friend abroad
But swallows, stirring to depart;
And my old shadow skimmed the road,
Happy without a heart.

Whether from those cold sunlit lands,
Or from the unimagined sky—
Blessed by some spirit's secret hands,
I drew the breath of sudden joy.

From all or from myself alone,
I asked not, being wise again;
But oh, the silver pastures shone,
And were translated then . . .

Fall not, soul, from those grave heights!
Look, to-day is bright with sun.
Hold back that fear; the darkest nights
See dawn, and soon are done.

FRANK KENDON (20th century)

Evening

Spare and like honey is the clovery texture
Of evening at pasture, the glint hush of summer
Falling in hueless folds,
Trailing at the horizon a cool hem;
Evening intact yet tremulous with far clamour
Of calling ewes and lambs—
Could love but fracture
One bone-grey sun-ray between dreamer
And dreamer!—
Calling until the wold's
Dun waters close invisibly over them.

LILIAN BOWES LYON (1895-1949)

These Fields at Evening

They wear their evening light as women wear
  Their pale, proud beauty for a lover's sake,
Too quiet-hearted evermore to care
  For moving worlds and musics that they make.
And they are hushed as lonely women are,
  So lost in dreams they have no thought to mark
How the wide heavens blossom, star by star,
  And the slow dusk is deepening to the dark.

The moon comes like a lover from the hill,
  Leaning across the twilight and the trees,
And finds them grave and beautiful and still,
  And wearing always, on such nights as these,
A glimmer less than any ghostly light,
  As women wear their beauty in the night.

DAVID MORTON (1886-1911)

from *To Autumn*

O Autumn, laden with fruit, and stainèd
With the blood of the grape, pass not, but sit
Beneath my shady roof; there thou may'st rest
And tune thy jolly voice to my fresh pipe,
And all the daughters of the year shall dance!
Sing now the lusty song of fruits and flowers . . .

. . . 'The Spirits of the Air live on the smells
Of fruit; and Joy, with pinions light, roves round
The gardens, or sits singing in the trees,'
Thus sang the jolly Autumn as he sat;
Then rose, girded himself, and o'er the bleak
Hills fled from our sight; but left his golden load.

WILLIAM BLAKE (1757-1827)

from *Strange Fruit*

This year the grain is heavy-ripe;
The apple shows a ruddier stripe;
Never berries so profuse
Blackened with so sweet a juice
On brambly hedges, summer-dyed.
The yellow leaves begin to glide;
But Earth in careless lap-ful treasures
Pledge of over-brimming measures,
As if some rich unwonted zest
Stirred prodigal within her breast . . .

LAURENCE BINYON (1869-1943)

Cock-Pheasant

Gilded with leaf-thick paint; a steady
Eye fixed like a ruby rock;
Across the cidrous banks of autumn
Swaggers the stamping pheasant-cock.

The thrusting nut and bursting apple
Accompany his jointed walk,
The creviced pumpkin and the marrow
Bend to his path on melting stalk.

Sure as an Inca priest or devil,
Feathers stroking down the corn,
He blinks the lively dust of daylight,
Blind to the hunter's powder-horn.

For me, alike, this flushed October—
Ripe, and round-fleshed, and bellyful—
Fevers me fast but cannot fright, though
Each dropped leaf shows the winter's skull.

LAURIE LEE (b 1914)

The Pheasant

Cock stubble-searching pheasant, delicate
Stepper, Cathayan bird, you fire
The landscape, as across the hollow lyre
Quick fingers burn the moment: call your mate
From the deep woods tonight, for your surprised
Metallic summons answers me like wire
Thrilling with messages, and I cannot wait
To catch its evening import, half-surmised.
Others may speak these things, but you alone
Fear never noise, make the damp thickets ring
With your assertions, set the afternoon
Alight with coloured pride. Your image glows
At autumn's centre—bright, unquestioning
Exotic bird, haunter of autumn hedgerows.

SIDNEY KEYES (1922-1943)

Autumnal Threads

'Twas when the fields had shed their golden grain
And burning suns had scar'd the russet plain;
No more the rose or hyacinth were seen,
Nor yellow cowslip on the tufted green:
But the rude thistle rear'd its hoary crown,
And the ripe nettle shew'd an irksome brown.
In mournful plight the tarnish'd groves appear,
And nature weeps for the declining year:
The sun, too quickly, reach'd the western sky,
And rising vapours hid his ev'ning eye:
Autumnal threads around the branches flew,
While the dry stubble drank the falling dew.

MARY LEAPOR (1722-1746)

The Latter Rain

The latter rain,—it falls in anxious haste
Upon the sun-dried fields and branches bare,
Loosening with searching drops the rigid waste,
As if it would each root's lost strength repair;
But not a blade grow green as in the spring,
No swelling twig puts forth its thickening leaves;
The robins only mid the harvests sing,
Pecking the grain that scatters from the sheaves;
The rain falls still,—the fruit all ripened drops,
It pierces chestnut burr and walnut shell,
The furrowed fields disclose the yellow crops,
Each bursting pod of talents used can tell,
And all that once received the early rain
Declare to man it was not sent in vain.

JONES VERY (1813-1880)

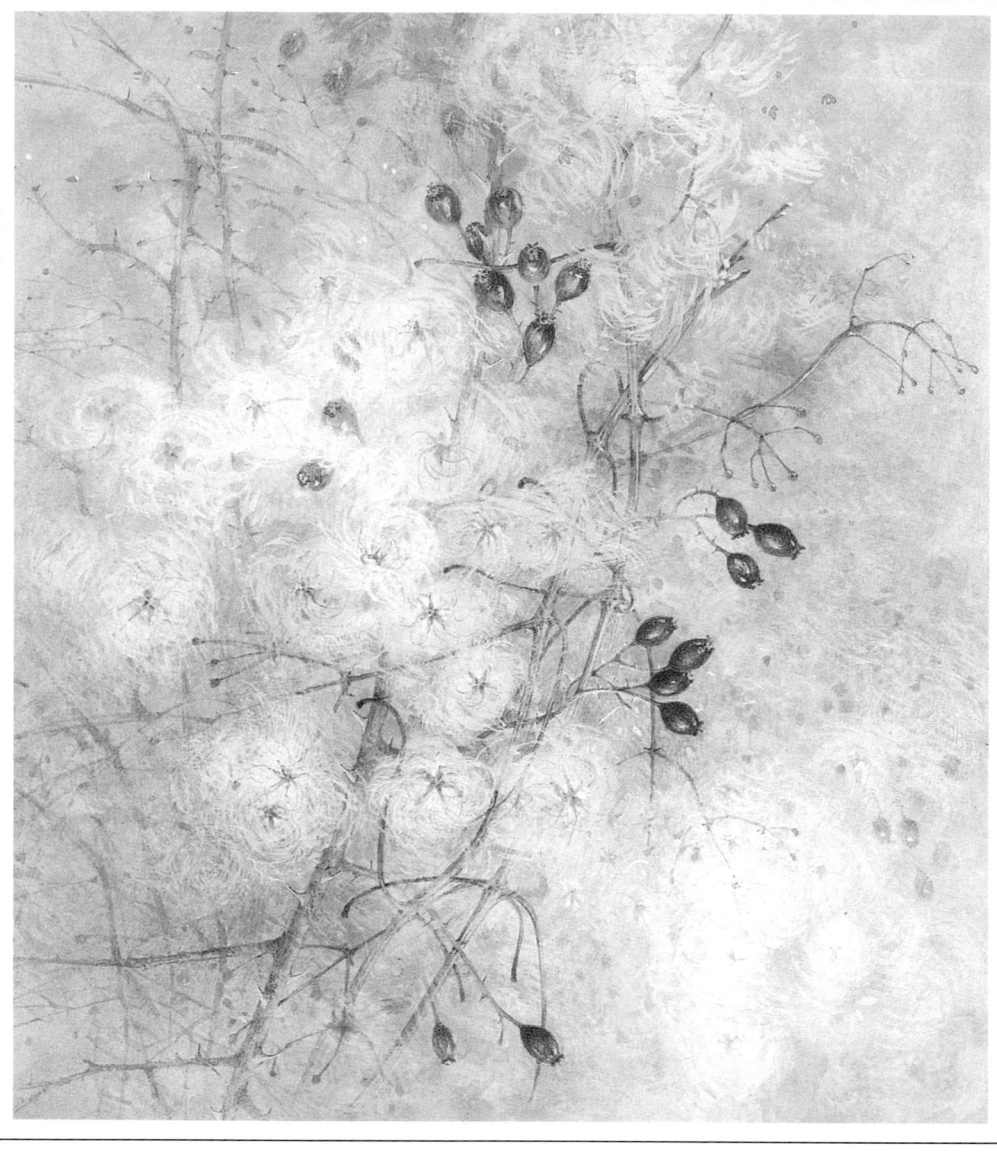

from *Wilderness*

The blackthorns hung with clinging sloes
Blue-veiled in weather coming cold,
And ruby-tasselled shepherd's-rose
Where flock the finches plumed with gold,
And swarming brambles laden still
Though boys and wasps have ate their fill.

EDMUND BLUNDEN (1896-1974)

Traveller's Joy

Through the valley and over the down
The withering hedge bends dry and brown,
The sycamore leaves hang rent and seared,
And the Traveller's Joy is Old Man's Beard—
Up the marsh and over the lea
The milk-white gulls sail up from the sea—
And it's O for the wind and the weeping rain,
And the summers that never shall rise again
Whatever may come to be.

ROSAMUND MARRIOTT WATSON (1863-1911)

from *Seed-Time*

Flowers of the willow-herb are wool;
Flowers of the briar berries red;
Speeding their seed as the breeze may rule,
Flowers of the thistle loosen the thread.
Flowers of the clematis drip in beard,
Slack from the fir-tree youngly climbed;
Chaplets in air, flies foliage seared;
Heeled upon earth, lie clusters rimed.

Now seems none but the spider lord;
Star in circle his web waits prey,
Silvering bush-mounds, blue brushing sward;
Slow runs the hour, swift flits the ray.
Now to his thread-shroud is he nigh,
Nigh to the tangle where wings are sealed,
He who frolicked the jewelled fly;
All is adroop on the down and the weald.

Verily now is our season of seed,
Now in our Autumn; and Earth discerns
Them that have served her in them that can read,
Glassing, where under the surface she burns,
Quick at her wheel, while the fuel, decay,
Brightens the fire of renewal: and we?
Death is the word of a bovine day,
Know you the breast of the springing To-be.

GEORGE MEREDITH (1828-1909)

St Luke's Summer

Now is the tolling time
Between the falling and the buried leaf;
A solitary bell
Saddens the soft air with the last knell
Of summer.
Gone is the swallow's flight, the curving sheaf;
The plums are bruised that hung from a bent bough,
Wasp-plundered apples in the dew-drenched grass
Lie rotting now.
Doomed with the rest, the daggered hawthorn bleeds
Bright crimson beads
For the birds' feast.
Gone are the clusters of ripe cherries,
Tart crabs and damsons where a bullfinch tarried,
Only the camp-fire coloured rowan berries
Blaze on.
Now is the time of slow, mist-hindered dawns,
Of sun that stains
Weeds tarnished early in the chilling rains,
Of coarse-cut stubble fields
Where starlings gather, busy with the scant grain,
And with hoarse chattering proclaim
The spent season.
Now are the last days of warm sun
That fires the rusted bracken on the hill,
And mellows the deserted trees
Where the last leaves cling, sapless, shrunk, and yellow.

A robin finds some warm October bough
Recapturing his song
Of Aprils gone,
And tardy blackbirds in the late-green larch
Remember March.

PHOEBE HESKETH (b 1909)

from *North Wind in October*

In the golden glade the chestnuts are fallen all;
From the sered boughs of the oak the acorns fall:
The beech scatters her ruddy fire;
The lime hath stripped to the cold,
And standeth naked above her yellow attire:
The larch thinneth her spire
To lay the ways of the wood with cloth of gold.

   Out of the golden green and white
Of the brake the fir trees stand upright
In the forest of flame, and wave aloft
To the blue of heaven their blue-green tuftings soft.

ROBERT BRIDGES (1844-1930)

A Windy Day

This wind brings all dead things to life,
Branches that lash the air like whips
And dead leaves rolling in a hurry
Or peering in a rabbits' bury
Or trying to push down a tree;
Gates that fly open to the wind
And close again behind,
And fields that are a flowing sea
And make the cattle look like ships;
Straws glistening and stiff
Lying on air as on a shelf
And pond that leaps to leave itself;
And feathers too that rise and float,
Each feather changed into a bird,
And line-hung sheets that crack and strain;
Even the sun-greened coat,
That through so many winds has served,
The scarecrow struggles to put on again.

ANDREW YOUNG (1885-1971)

Autumn

Therefore their latter journey to the grave
Was like those days of later autumn tide
When he who in some town may chance to bide
Opens the window for the balmy air,
And seeing the golden hazy sky so fair,
And from some city garden hearing still
The wheeling rooks the air with music fill,
Sweet hopeful music, thinketh: Is this Spring?
Surely the year can scarce be perishing?
But then he leaves the clamour of the town,
And sees the scanty withered leaves fall down,
The half-ploughed field, the flowerless garden plot,
The dark full stream by summer long forgot,
The tangled hedges where, relaxed and dead,
The twining plants their withered berries shed,
And feels therewith the treachery of the sun,
And knows the pleasant time is well nigh done.

WILLIAM MORRIS (1834-1896)

November

The mellow year is hasting to its close;
The little birds have almost sung their last,
Their small notes twitter in the dreary blast—
That shrill-piped harbinger of early snows:
The patient beauty of the scentless rose,
Oft with the Morn's hoar crystal quaintly glass'd,
Hangs, a pale mourner for the summer past,
And makes a little summer where it grows:
In the chill sunbeam of the faint brief day
The dusky waters shudder as they shine,
The russet leaves obstruct the straggling way
Of oozy brooks, which no deep banks define,
And the gaunt woods, in ragged, scant array,
Wrap their old limbs with sombre ivy twine.

HARTLEY COLERIDGE (1796-1849)

The Last Week in October

The trees are undressing, and fling in many places—
On the gray road, the roof, the window-sill—
Their radiant robes and ribbons and yellow laces;
A leaf each second so is flung at will,
Here, there, another and another, still and still.

A spider's web has caught one while downcoming,
That stays there dangling when the rest pass on;
Like a suspended criminal hangs he, murmuring
In golden garb, while one yet green, high yon,
Trembles, as fearing such a fate for himself anon.

THOMAS HARDY (1840-1928)

from *The Storm is Over*

But ah! the leaves of summer that lie on the ground!
What havoc! The laughing timbrels of June,
That curtained the birds' cradles, and screened their song,
That sheltered the cooing doves at noon,
Of airy fans the delicate throng,—
Torn and scattered around:
Far out afield they lie,
In the watery furrows die,
In grassy pools of the flood they sink and drown,
Green-golden, orange, vermilion, golden and brown,
The high year's flaunting crown
Shattered and trampled down.

ROBERT BRIDGES (1844-1930)

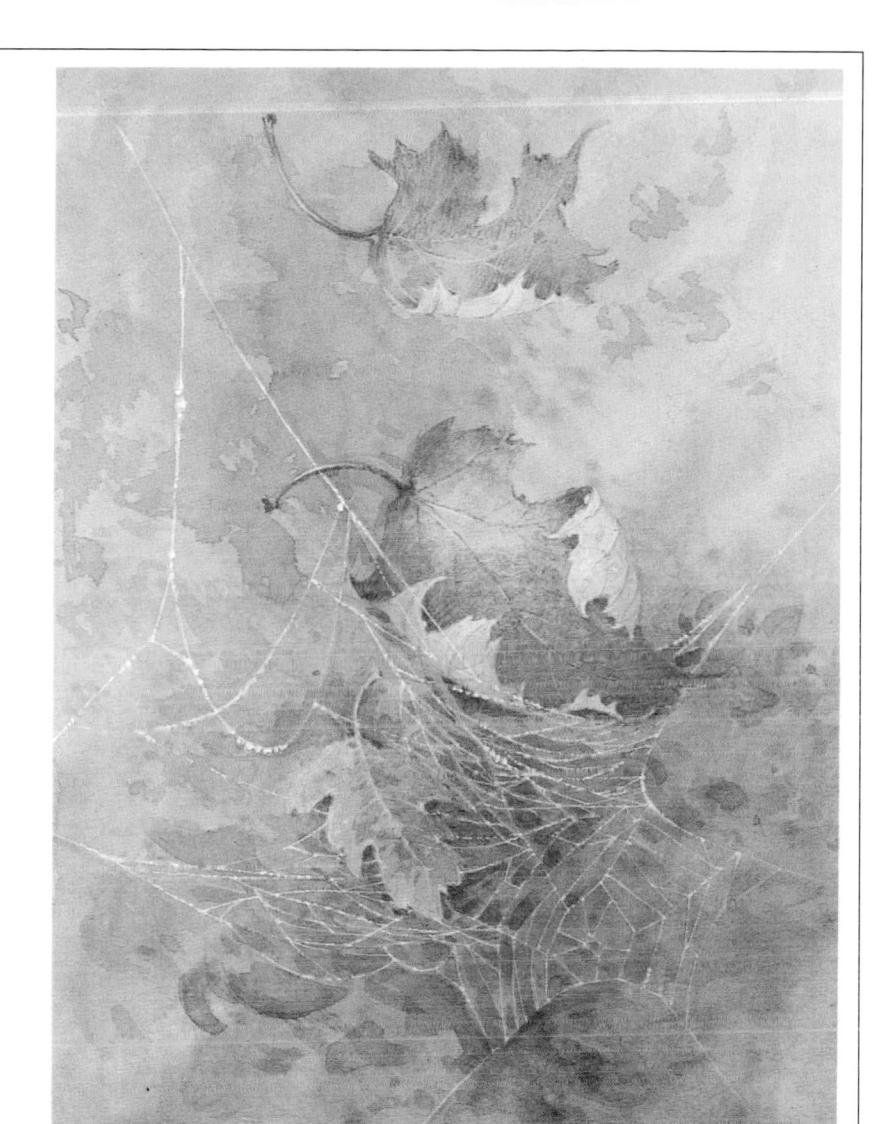

Last Hours

A gray day and quiet,
With slow clouds of gray,
And in dull air a cloud that falls, falls
All day.

The naked and stiff branches
Of oak, elm, thorn,
In the cold light are like men aged and
Forlorn.

Only a gray sky,
Grass, trees, grass again,
And all the air a cloud that drips, drips,
All day.

Lovely the lonely
Bare trees and green grass—
Lovelier now the last hours of slow winter
Slowly pass.

JOHN FREEMAN (1880-1929)

November Skies

Than these November skies
Is no sky lovelier. The clouds are deep;
Into their gray the subtle spies
Of colour creep,
Changing that high austerity to delight,
Till even the leaden interfolds are bright.
And, where the cloud breaks, faint far azure peers
Ere a thin flushing cloud again
Shuts up that loveliness, or shares.
The huge great clouds move slowly, gently, as
Reluctant the quick sun should shine in vain,
Holding in bright caprice their rain.
And when of colours none,
Not rose, nor amber, nor the scarce late green,
Is truly seen,—
In all the myriad gray,
In silver height and dusky deep, remain
The loveliest,
Faint purple flushes of the unvanquished sun.

JOHN FREEMAN (1880-1929)

November Night

Listen.
With faint-dry sound,
Like steps of passing ghosts,
The leaves, frost-crisp'd, break from the trees
And fall.

ADELAIDE CRAPSEY (1878-1914)

What is Winter?

What is winter? A word,
A figure, a clever guess.
That time-word does not answer to
This drowsy wakefulness.
The secret stream scorns the interval
Though the calendar shouts one from the wall;
The spirit has no last days;
And death is no more dead than this
Flower-haunted haze.

EDMUND BLUNDEN (1896-1974)

Reluctance

Out through the fields and the woods
　And over the walls I have wended;
I have climbed the hills of view
　And looked at the world, and descended;
I have come by the highway home,
　And lo, it is ended.

The leaves are all dead on the ground,
　Save those that the oak is keeping
To ravel them one by one
　And let them go scraping and creeping
Out over the crusted snow,
　When others are sleeping.

And the dead leaves lie huddled and still,
　No longer blown hither and thither;
The last lone aster is gone;
　The flowers of the witch-hazel wither;
The heart is still aching to seek,
　But the feet question 'Whither?'

Ah, when to the heart of man
　Was it ever less than a treason
To go with the drift of things,
　To yield with a grace to reason,
And bow and accept the end
　Of a love or a season?

ROBERT FROST (1874-1963)

Christmas

I had almost forgotten the singing in the streets,
Snow piled up by the houses, drifting
Underneath the door into the warm room,
Firelight, lamplight, the little lame cat
Dreaming in soft sleep on the hearth, mother dozing,
Waiting for Christmas to come, the boys and me
Trudging over blanket fields waving lanterns to the sky.
I had almost forgotten the smell, the feel of it all,
The coming back home, with girls laughing like stars,
Their cheeks, holly berries, me kissing one,
Silent-tongued, soberly, by the long church wall;
Then back to the kitchen table, supper on the white cloth,
Cheese, bread, the home-made wine;
Symbols of the night's joy, a holy feast.
And I wonder now, years gone, mother gone,
The boys and girls scattered, drifted away with the snowflakes,
Lamplight done, firelight over,
If the sounds of our singing in the streets are still there,
Those old tunes, still praising;
And now, a life-time of Decembers away from it all,
A branch of remembering holly spears my cheeks,
And I think it may be so;
Yes, I believe it may be so.

LEONARD CLARK (1905-1981)

Holly

'Tis a brave tree. While round its boughs in vain
The warring wind of January bites and girds,
It holds the clusters of its crimson grain,
A winter pasture for the shivering birds.
Oh, patient holly, that the children love,
No need for thee of smooth blue skies above:
Oh, green strong holly, shine amid the frost;
Thou dost not lose one leaf for sunshine lost.

AUGUSTA WEBSTER (1837-1894)

from *The Setting Sun*

Then comes the Winter, like a hale old man
Wrapped in his cloak with frosty locks and beard.
Winter is the time for clear cold starlight nights,
And driving snows, and frozen roads and rivers,
For crowding round the blazing Christmas fire,
For telling tales that make the blood run cold,
For sipping elder-wine and cracking filberts,
For friendships, chilblains, fun, roast beef, mince pies,
And shivering fits on jumping into bed:
And thus the year goes round, and round, and round.

JAMES HURNARD (late 19th-early 20th century)

Blow, Blow, Thou Winter Wind
from *As You Like It*

Blow, blow, thou winter wind,
Thou are not so unkind
  As man's ingratitude;
Thy tooth is not so keen,
Because thou art not seen,
  Although thy breath be rude.

Heigh ho! sing, heigh ho! unto the green holly:
Most friendship is feigning, most loving mere folly:
  Then, heigh ho, the holly!
  This life is most jolly.

  Freeze, freeze, thou bitter sky,
  That dost not bite so nigh
    As benefits forgot:
  Though thou the waters warp,
  Thy sting is not so sharp
    As friend remembered not.

Heigh ho! sing, heigh ho! &c.

WILLIAM SHAKESPEARE (1564-1616)

Now stir the fire, and close the shutters fast,
Let fall the curtains, wheel the sofa round,
And, while the bubbling and loud-hissing urn
Throws up a steamy column, and the cups,
That cheer but not inebriate, wait on each,
So let us welcome peaceful evening in . . .

  Oh Winter, ruler of th'inverted year, . . .
I love thee, all unlovely as thou seem'st,
And dreaded as thou art. Thou hold'st the sun
A prisoner in the yet undawning east,
Shortening his journey between morn and noon,
And hurrying him, impatient of his stay,
Down to the rosy west; but kindly still
Compensating his loss with added hours
Of social converse and instructive ease,
And gathering, at short notice, in one group
The family dispersed, and fixing thought,
Not less dispersed by day-light and its cares.
I crown thee king of intimate delights,
Fire-side enjoyments, home-born happiness,
And all the comforts that the lowly roof
Of undisturbed retirement and the hours
Of long uninterrupted evening know.

WILLIAM COWPER (1731-1800)

from *Frost at Midnight*

Therefore all seasons shall be sweet to thee,
Whether the summer clothe the general earth
With greenness, or the redbreast sit and sing
Betwixt the tufts of snow on the bare branch
Of mossy apple-tree, while the nigh thatch
Smokes in the sun-thaw; whether the eave-drops fall
Heard only in the trances of the blast,
Or if the secret ministry of frost
Shall hang them up in silent icicles,
Quietly shining to the quiet Moon.

SAMUEL TAYLOR COLERIDGE (1779-1834)